CURRICULUM AND EVALUATION

S T A N D A R D S

FOR SCHOOL MATHEMATICS

ADDENDA SERIES, GRADES 5–8

# DEVELOPING NUMBER SENSE IN THE MIDDLE GRADES


Barbara J. Reys

*with*

Rita Barger, Barbara Dougherty, Linda Lembke, Andy Parnas, Ruthi Sturdevant, Maxim Bruckheimer, Jack Hope, Zvia Markovits, Sue Reehm, Marianne Weber

Consultants

Diana Lambdin Kroll, Frank K. Lester, Jr., Kenneth P. Goldberg

Frances R. Curcio, Series Editor



NATIONAL COUNCIL OF TEACHERS OF MATHEMATICS

*Second Printing 1992*

Library of Congress Cataloging-in-Publication Data:

Reys, Barbara.
Developing number sense in the middle grades / Barbara Reys, with Rita Barger [et al.].
p. cm.
Includes bibliographical references.
ISBN 0-87353-322-4
1. Number concept in children—Study and teaching (Elementary)
I. Barger, Rita. II. Title.
QA141.15.R48 1991
372.7—dc20 91-12623
CIP



Printed in the United States of America

# TABLE OF CONTENTS

## FOREWORD

In March 1989, the National Council of Teachers of Mathematics officially released the *Curriculum and Evaluation Standards for School Mathematics* (NCTM 1989). The document provides a vision and a framework for strengthening the mathematics curriculum in kindergarten to grade 12 in North American schools. Evaluation, an integral part of planning for and implementing instruction, is an important feature of the document. Also presented is a contrast between traditional rote methods of teaching, which have proven to be unsuccessful, and recommendations for improving instruction supported by current educational research.

As the *Curriculum and Evaluation Standards* was being developed, it became apparent that a plethora of examples would be needed to illustrate how the vision could be realistically implemented in the K–12 classroom. A Task Force on the Addenda to the *Curriculum and Evaluation Standards for School Mathematics,* chaired by Thomas Rowan and composed of Joan Duea, Christian Hirsch, Marie Jernigan, and Richard Lodholz, was appointed by Shirley Frye, then NCTM president, in the spring of 1988. The Task Force's recommendations on the scope and nature of the supporting publications were submitted in the fall of 1988 to the Educational Materials Committee, which subsequently framed the Addenda Project for NCTM Board approval. Following the release of the *Curriculum and Evaluation Standards* and on the basis of recommendations of the NCTM Task Force (Rowan 1988), writing teams were assigned to develop addenda to provide teachers with classroom ideas for translating the *Standards* into classroom practice. Three writing teams were formed to prepare materials for grades K–6 (Miriam A. Leiva, editor), grades 5–8 (Frances R. Curcio, editor), and grades 9–12 (Christian R. Hirsch, editor).

The themes of problem solving, reasoning, communication, connections, technology, and evaluation have been woven throughout the materials. The writing teams included K–12 classroom teachers, supervisors, and university mathematics educators. The materials have been field tested in an effort to make them "teacher friendly."

Implementing the *Curriculum and Evaluation Standards* will take time. The Addenda Series was written to help teachers as they undertake this task. Furthermore, the Addenda Series is appropriate for use in inservice staff development as well as for use in preservice courses in teacher education programs.

On behalf of the National Council of Teachers of Mathematics, I would like to thank all the authors, consultants, and editors who gave willingly of their time, effort, and expertise in developing these exemplary materials. In particular, I would like to acknowledge gratefully the work of Barbara J. Reys and her contributing authors, Rita Barger, Maxim Bruckheimer, Barbara Dougherty, Jack Hope, Linda Lembke, Zvia Markovits, Andy Parnas, Sue Reehm, Ruthi Sturdevant, and Marianne Weber for preparing the *Developing Number Sense in the Middle Grades* manuscript. Special thanks are extended to Diana Lambdin Kroll and Frank K. Lester, Jr., evaluation consultants, and Kenneth P. Goldberg, technology consultant, for their contributions to the entire Grades 5–8 Addenda Series. Finally, this project would not have materialized without the outstanding technical support supplied by Cynthia Rosso and the NCTM staff.

Bonnie H. Litwiller, *Addenda Project Coordinator*

## *PREFACE*

The purpose of *Developing Number Sense in the Middle Grades*, as well as of the other books in the Grades 5–8 Addenda Series, is to provide teachers with ideas and materials to support the implementation of the *Curriculum and Evaluation Standards for School Mathematics* (NCTM 1989). In addition to this book, the four other publications in this series are *Understanding Rational Numbers and Proportions* (Bezuk, forthcoming), *Geometry in the Middle Grades* (Geddes 1992), *Patterns and Functions* (Phillips 1991), and *Dealing with Data and Chance* (Zawojewski 1991). These books are *not* an outline of a middle school curriculum, but rather, they present several topics and activities to exemplify the ideas advocated in the *Curriculum and Evaluation Standards,* and they provide examples to help students make the transition from elementary to high school mathematics.

The unifying themes of the *Curriculum and Evaluation Standards*, the characteristics of a new classroom learning environment, and the role of evaluation are described below. The discussion of each topic will refer to examples in this book.

### *Unifying Themes*

The unifying themes of the *Curriculum and Evaluation Standards* include mathematics as problem solving, mathematics as communication, mathematical reasoning, and mathematical connections. These themes are not separate, isolated entities, but rather, they are all interrelated. The ideas, examples, illustrations, and activities presented in this book were designed to demonstrate the interrelationships by weaving these themes throughout the activities as well as by providing ideas for incorporating technology and evaluation techniques.

*Mathematics as problem solving.* Although problem solving has been a goal of mathematics instruction throughout the years, it became the focus of attention with the advent of NCTM's *An Agenda for Action* (NCTM 1980). The *Curriculum and Evaluation Standards* reaffirms the importance of problem solving in mathematics instruction.

The excitement of learning and applying mathematics is generated when problems develop within the context of a situation familiar to students. Allowing them to formulate problems as they naturally arise within the context of everyday experiences gives them the opportunity to put mathematics to work, observing its usefulness and its applicability (e.g., Activity 1). However, "not all problems require a real-world setting. Indeed, middle school students often are intrigued by story settings or those arising from mathematics itself" (NCTM 1989, p. 77). For example, see Activities 4 and 8.

*Mathematics as communication.* Although the formal language of mathematics is concise, high in concept density, and may be seemingly foreign, students should have the opportunity to bring meaning to mathematics based on their prior knowledge and background experiences. Allowing them to talk about their experiences and how they relate to mathematics concepts, to listen to each other as they share ideas, to read mathematics in various formats (e.g., number sentences, graphs, charts), and to write about mathematical situations affords students the opportunity to compare experiences, clarify their thinking, and develop an understanding of how the mathematics they study in school is related to the mathematics they experience in the "real world." This requires the integration of the four language arts—speaking, listening, reading, and writing—with the mathematics lesson (e.g., Activity 2).

Communicating in mathematics requires a common language and familiarity with modes of representing mathematical ideas. Depending on students' "comfort level," oral language, prose, manipulatives, pictures, diagrams, charts, graphs, or symbols can be used to communicate ideas (e.g., Activities 8 and 43). Students should be encouraged to translate from one mode to another.

*Mathematics as reasoning.* During the middle school years, students should have opportunities to develop and employ their abilities in logical and spatial reasoning, as well as in proportional and graphical reasoning. The development of a student's reasoning ability occurs over a period of time. One can observe extreme differences between students in grade 5 and students in grade 8. As a result, instructional approaches must reflect these differences.

Depending on students' readiness, exploratory activities, experiments, and projects may require them to give a descriptive account of what they observe, an informal argument based on empirical results, or a formal proof to support a conjecture (Hirsch and Lappan 1989). Active learners should be constantly involved in questioning, examining, conjecturing, and experimenting (e.g., Activity 3).

*Mathematical connections.* Traditionally, mathematics has often been presented as an isolated set of rules to be memorized. The *Curriculum and Evaluation Standards* suggests that mathematics be presented as an integrated whole. Students should observe the interrelatedness among branches of mathematics: number theory, geometry, algebra, and so on (e.g., Activity 8). Furthermore, students should become aware of how mathematics is related to other disciplines such as science, art, literature, music, social studies, business, and industrial technology (e.g., Activity 1).

There is no guarantee that allowing students to explore, create, and experiment within the context of a problem-solving setting will lead them to discover connections between and among mathematics concepts. Teachers may need to guide students in discovering connections and to elicit these connections explicitly. Students who recognize connections within mathematics and with other disciplines can understand and appreciate the logical unity and the power of mathematics (Steen 1989).

#### A New Classroom Environment

Implementing the *Curriculum and Evaluation Standards* requires a new way of teaching. The traditional teacher roles of authority figure and information disseminator must change to learning facilitator and instructional decision maker.

Knowledge about students and how they learn mathematics can contribute to establishing a conducive learning environment for middle school students. The teacher selects the instructional objectives on the basis of knowledge of his or her students, knowledge of mathematics, and knowledge of pedagogy (NCTM 1991). After selecting the instructional objectives, the teacher must decide how to deliver the content. Is the use of manipulatives appropriate? Is the use of technology appropriate? Is a cooperative learning setting appropriate?

*Appropriate use of manipulatives.* Manipulatives are multisensory tools for learning that provide students with a means to communicate ideas by allowing them to model or represent their ideas concretely. Using manipulatives, however, does not guarantee the understanding of a mathematics concept (Baroody 1989). After allowing students to explore using manipulatives, teachers must formulate questions to elicit the important, "big" mathematical ideas that enable students to make connections between the mathematics and the manipulatives used to represent the concept(s) (e.g., Activities 17 and 18).

*Appropriate use of technology.* Developments in technology have made the traditional, computation-dominated mathematics curriculum obsolete. As a result, technology has been given a prominent place in the *Curriculum and Evaluation Standards*, in terms of which technology should be made available for use in the classroom and how technology should be used in mathematics instruction. It is expected that at the middle school level, students will have access to appropriate calculators; computers should be available for demonstration purposes as well as for individual and group work (NCTM 1989, p. 8). It should be noted that new advances in technology are being made constantly. Teachers should keep abreast of new developments that support mathematics instruction.

In this book, suggestions are made for integrating the use of calculators to explore relative sizes of numbers (e.g., Activities 4 and 26), facilitate and expedite computation (e.g., Activities 8 and 12), and generate examples to verify students' thinking (e.g., Activity 25). Specifically, simple four-function calculators and calculators with fraction capability are appropriate for use in many activities.

Integrating the use of the computer is also recommended. In particular, depending on their abilities and background experiences, students may enter a simple BASIC program to demonstrate computer counting, or they can create a program to count (e.g., Activity 9).

*Appropriate use of cooperative learning groups.* Traditionally, mathematics has been taught as a "solo," isolated activity, yet in business and industry, mathematicians often work in teams to solve problems and attain common objectives (Steen 1989). Allowing students to work in cooperative groups affords them the opportunity to develop social and communication skills while working with peers of various ethnic, religious, and racial groups.

Cooperative learning environments, characterized by students working together and interacting with each other, contribute to internalizing concepts by forcing students to defend their views against challenges brought by their peers. The value of this approach is supported by the work of Vygotsky ([1934]1986) who discussed the increasingly interrelated nature of language and cognition as children grow.

Cooperative groups usually contain three to five students and may be established for various lengths of time (Artzt and Newman 1990; Davidson 1989). Unlike most traditional small-group instruction in reading and mathematics, cooperative learning groups are heterogeneous and everyone must work together for the common good of all. Students who understand the concept being discussed are responsible for explaining it to those who do not understand. When using cooperative groups, teachers must consider new ways of evaluating performance to ensure the success of the instructional objective(s).

#### *The Role of Evaluation*

Making changes in the content and methods of mathematics instruction will also require making changes in why and how students' work is evaluated. Evaluation should be an integral part of instruction, not limited to grading and testing. According to Kroll and Lester (see Appendix 1), there are at least four reasons for collecting evaluation information:

- to make decisions about the content and methods of mathematics instruction (e.g., Activity 11);
- to make decisions about classroom climate (e.g., p. 12);
- to help in communicating what is important (e.g., Activity 15);
- to assign grades (e.g., Activity 39).

In other words, evaluation includes much more than marking right and wrong answers. It "must be more than testing; it must be a continuous, dynamic, and often informal process" (NCTM 1989, p. 203). What methods can be used for evaluation purposes? The *Curriculum and Evaluation Standards* recommends that teachers use a variety of types of evaluation. In Appendix 1, four major categories of evaluation are discussed. Examples can be found throughout the text as indicated: (1) *observing and questioning students* (e.g., Activity 11), (2) *using assessment data reported by students* (e.g., p. 6), (3) *assessing students' written mathematics work* (e.g., Activity 22), and (4) *using multiple-choice or short-answer items* (e.g., Activity 2). Teachers might, for example, observe and question students to assess their understanding and to gain insight into their feelings and their beliefs about mathematics; use holistic scoring techniques for a focused assessment of students' written problem-solving work; or collect information through students' responses to short-answer questionnaires or through written assignments such as journal entries or brief essays. These and other evaluation techniques are discussed in more detail in Appendix 1.

This brief description of the unifying themes of the *Curriculum and Evaluation Standards,* the characteristics of a new learning environment, and the role of evaluation is furnished as a starting point in understanding and appreciating the ideas presented in this book. It is hoped that these ideas will provide a foundation for developing number sense in the middle grades and will generate interest among teachers for improving instructional and evaluation techniques in mathematics.

Frances R. Curcio, Editor
Grades 5–8 Addenda Series

◆ ◆ ◆ ◆ ◆ ◆ ◆ ◆

## INTRODUCTION

> Within a unit on fractions, a sixth-grade teacher asked her students to tell as many things as they could about the fraction $\frac{2}{5}$. The students made statements such as these: It's about the same as a half. It's smaller than $\frac{1}{2}$. It's bigger than $\frac{1}{4}$. It's bigger than $\frac{1}{3}$. It's twice as large as $\frac{1}{5}$. Two $\frac{2}{5}$'s would almost make a whole. It can be named as 0.4. It's the same as 40 percent. It's a proper fraction. It's a rational number. It's the same as $\frac{8}{20}$. Monday and Tuesday make $\frac{2}{5}$ of our school week.

The students in this class are displaying a good sense for fractions—relating them to their own experiences, creating extensions of those experiences, and exhibiting a sense for the size, meaning, and variety of expressions associated with the fraction $\frac{2}{5}$. The teacher is encouraging the development of this "number sense" by posing a question and allowing time for the students to collect and verbalize their knowledge, to listen to each other and build on each other's ideas, and to create their own meaningful representations.

Within the past few years, growing attention has been directed toward what is now being called *number sense*. The entire February 1989 issue of the *Arithmetic Teacher* was devoted to the topic; a conference for researchers, "Establishing Foundations for Research on Number Sense and Related Topics," was sponsored by the National Science Foundation in 1989; *Everybody Counts,* a recent publication of the National Research Council, states: "The major objective of elementary school mathematics should be to develop number sense" (National Research Council 1989, p. 46). Finally, the *Curriculum and Evaluation Standards for School Mathematics* (NCTM 1989) includes number sense as a major theme throughout its recommendations:

> Understanding multiple representations for numbers is a crucial precursor to solving many of the problems students encounter. (p. 87)
>
> Students should analyze the way the various algorithms work and how they relate to the meaning of the operation and to the numbers involved. (p. 95)
>
> Children must understand numbers if they are to make sense of the ways numbers are used in their everyday world. (p. 38)
>
> Intuition about number relationships helps children make judgments about the reasonableness of computational results and of proposed solutions to numerical problems. (p. 38)

The *Curriculum and Evaluation Standards* highlights the importance of number sense for grades 5 to 8 in Standard 5: Number and Number Relationships and in Standard 7: Computation and Estimation (fig. 1). Furthermore, the changes being recommended with regard to computation focus on the need to develop number sense. Specifically,

> the greatest revisions to be made in the teaching of computation include the following: fostering a solid understanding of, and proficiency with, simple calculations; abandoning the teaching of tedious calculations using paper-and-pencil algorithms in favor of exploring more mathematics; fostering the use of a wide variety of computation and estimation techniques—ranging from quick mental calculation to those using computers—suited to different mathematical settings; developing the skills necessary to use appropriate technology and then translating computed results to the problem setting; [and] providing students with ways to check the reasonableness of computations (number and algorithmic sense, estimation skills). (NCTM 1989, p. 95)

Standard 5: Number and Number Relationships

In grades 5–8, the mathematics curriculum should include the continued development of number and number relationships so that students can—

- understand, represent, and use numbers in a variety of equivalent forms (integer, fraction, decimal, percent, exponential, and scientific notation) in real-world and mathematical problem situations;
- develop number sense for whole numbers, fractions, decimals, integers, and rational numbers;
- understand and apply ratios, proportions, and percents in a wide variety of situations;
- investigate relationships among fractions, decimals, and percents;
- represent numerical relationships in one- and two-dimensional graphs.

(NCTM 1989, pp. 87–90)

Standard 7: Computation and Estimation

In grades 5–8, the mathematics curriculum should develop the concepts underlying computation and estimation in various contexts so that students can—

- compute with whole numbers, fractions, decimals, integers, and rational numbers;
- develop, analyze, and explain procedures for computation and techniques for estimation;
- develop, analyze, and explain methods for solving proportions;
- select and use an appropriate method for computing from among mental arithmetic, paper-and-pencil, calculator, and computer methods;
- use computation, estimation, and proportions to solve problems;
- use estimation to check the reasonableness of results.

(NCTM 1989, pp. 94–97)

***Fig. 1***

Number sense is considered a desirable trait to foster, but its meaning, like that of common sense, is difficult to express simply. It is easier to depict situations in which students either do or do not exhibit number sense than it is to state a clear definition of the term. Consider the following examples:

> When asked to find the product of 48 and 0.5 mentally, one student converted 0.5 to ½, then took ½ of 48 to get an answer of 24. In contrast, another student mentally "wrote" 48 above 0.5 in his mind and multiplied 5 times 8, remembering the 0 and "carrying" the 4. Then he took 5 times 4, getting 20, and added the 4 he carried, getting 24. Counting the number of decimal places in his problem, he knew to insert the decimal point in his answer between the 4 and 0, thus arriving at the answer, 24.0 or 24.

The first student used her knowledge about numbers and their relationships to solve the problem efficiently, whereas the second student merely applied a learned algorithm, which was not convenient to use mentally.

> Two students who had just been introduced to a protractor were asked to measure an angle. One recorded a measure of 30 degrees and the other recorded a measure of 150 degrees. In discussing who was correct, they determined that the angle was an acute angle; therefore, 30 degrees was the only sensible answer.

Students often make this 30°-150° mistake when first using protractors, not understanding which scale to read. They apply a set of procedures mechanically, not noticing that the answer isn't reasonable in light of the actual size of the angle pictured. Making connections from what is known (properties of acute angles) to new situations (using a protractor) is an important indicator of number sense (see Activity 41 for an illustration of this approach).

> In computing the product of 4.5 and 1.2, a student carefully lined up the decimals and then multiplied, bringing the decimal point straight down and reporting a product of 54.0 (fig. 2).

**Fig. 2**

Reflection on the answer should have caused the student to realize the product was too big. Multiplying 4.5 by a number slightly more than 1 produces an answer a little more than 4.5. Instead, this student applied an incorrect procedure (line up the decimals in the factors and bring the decimal point straight down) and did not reflect on whether the resulting answer was reasonable.

> Two students were using their calculators to figure sales tax on a set of items. For one problem, tax was calculated to be $54 on a $12 purchase.

The failure to enter the decimal form of the tax rate (4.5 percent) yielded a totally unreasonable answer. These students never stopped to question their answer.

These and similar situations are probably familiar to most middle school teachers. Although number sense may be a new term to many teachers, witnessing number sense, or the lack thereof, is not a new experience.

Clearly, teaching for number sense is not the same as teaching an algorithm. It cannot be accomplished by lecture, drill, and homework. Number sense, however, is not simply another topic to be squeezed into an already full curriculum. Teaching number sense is, instead, a change in approach to the teaching and learning of mathematics. It focuses on students and their solution strategies rather than on a "right answer"; on thinking rather than on the mechanical application of rules; and on student-generated solutions rather than on teacher-supplied answers.

This book addresses the issues and concerns of number sense. It looks at what number sense is—is it a new topic or a new perspective on teaching mathematics? It considers the teacher's role in fostering the development of number sense and provides examples and activities that illustrate teaching *for* number sense.

### *WHAT IS NUMBER SENSE?*

Phrases such as "number sense," "operation sense," and "intuitive understanding of number" are used throughout the *Curriculum and Evaluation Standards* to describe an intangible quality possessed by successful mathematicians. Some people would equate these terms with the general notion of developing conceptual understanding. They have been used to describe mental computation, basic fact acquisition, problem solving, higher-order thinking, reasonableness, and estimation. In fact, number sense plays a role in each of these areas; however, it is important to understand that none of these, singly or as a group, completely describes it.

Number sense refers to an intuitive feeling for numbers and their various uses and interpretations; an appreciation for various levels of accuracy when figuring; the ability to detect arithmetical errors; and a common-sense approach to using numbers. Number sense is not a finite entity that a student either has or does not have, nor is it a unit that can be

> ***Number sense builds on arithmetic as words build on the alphabet. Numbers arise in measurement, in chance, in data, and in geometry, as well as in arithmetic. Mathematics in elementary school should weave all these threads together to create in children a robust sense of numbers. (National Research Council 1989, p. 46)***

"taught," then put aside. It is also clear that number sense does not develop by chance, nor does manipulating numbers necessarily reflect this acquaintance and familiarity with numbers. Above all, number sense is characterized by a desire to make sense of numerical situations. Number sense is a way of thinking that must permeate all aspects of mathematics teaching and learning if mathematics is to make sense.

Although number sense is not easy to define, there is general agreement on two points: (1) in the same way that problem solving is the underlying theme for the study of mathematics, number sense is the underlying theme for the study of number; and (2) sense-making is the highest priority as numerical concepts are being constructed by the learner.

The idea of number sense is not new. As early as the 1930s William Brownell was concerned with what he called "meaningful learning." To Brownell the true test of mathematical learning was not the ability to compute but was the possession of an "intelligent grasp [of] number relations and the ability to deal with arithmetical situations with proper comprehension of their mathematical as well as their practical significance" (1972, p. 39). He stressed the importance of having students understand the meaning of all types of numbers, the meaning of the number system itself, and the meaning of place value, as well as the importance of the functions of the basic operations and the rationale of computation.

Number sense as used in the *Curriculum and Evaluation Standards* is reminiscent of Brownell's ideas and is characterized in the *Standards* in the following ways: (1) understanding the meanings of numbers, (2) having an awareness of multiple relationships among numbers, (3) recognizing the relative magnitude of numbers, (4) knowing the relative effect of operating on numbers, and (5) possessing referents for measures of common objects and situations in the environment (NCTM 1989, p. 38).

Perhaps number sense, like common sense, is best described by looking at the specific behavioral characteristics of those who possess a high degree of it. The student with number sense will look at a problem holistically before confronting the details of the problem. For example, in adding $1\frac{2}{3} + \frac{3}{4} + \frac{1}{3}$, the student might reorder the addends to $1\frac{2}{3} + \frac{1}{3} + \frac{3}{4}$ and mentally compute the sum of $2\frac{3}{4}$. The student will look for relationships among the numbers and operations and will consider the context in which the question was posed. Students with number sense will choose or even invent a method that takes advantage of their own understanding of the relationships between numbers and between numbers and operations, and they will seek the most efficient representation for the given task.

Number sense can also be recognized in the student's use of benchmarks to judge number magnitude (e.g., $\frac{2}{5}$ of 49 is less than half of 49), to recognize unreasonable results for calculations, and to employ nonstandard algorithms for mental computation and estimation. Other possible indicators of number sense have been highlighted by Resnick (in Sowder and Schappelle 1989) and include the following:

- using well-known number facts to figure out facts for which one is not so sure;
- approximating numerical answers when approximations are adequate and efficient;
- making sense of situations involving number and quantity; and

- using the decimal structure of the number system to decompose and recompose numbers to simplify calculations.

Number sense, then, is not another new topic for teachers to include in an already overcrowded mathematics program. Rather it is a different perspective from which to view the learning of mathematics. It is what Brownell called meaningful learning. *Number sense is both the ability of the learner to make logical connections between new information and previously acquired knowledge and the drive within the learner to make forming these connections a priority.* Number sense will be valued among students only if teachers believe that it is more important for students to make sense of the mathematics than to master rules and algorithms, which are often poorly understood. Teachers must also act on these beliefs. Students will acquire a good sense of number if they are engaged in purposeful activities requiring them to think about numbers and number relationships and to make connections to quantitative information seen in everyday life.

### THE TEACHER'S ROLE IN DEVELOPING NUMBER SENSE

Just as number sense requires a certain way of thinking, teaching for the development of number sense requires a certain approach to teaching. Rather than being an individual topic that can be addressed in a unit of instruction, promotion of number sense must be a part of all mathematics teaching and an underlying theme of mathematics as a sense-making activity. Teachers play an important role in building number sense by the type of classroom environment they create, by the teaching practices they employ, and by the activities they select.

***No substitute exists for a skillful teacher and an environment that fosters curiosity and exploration. (Howden 1989, p. 11)***

#### Create a Classroom Environment That Nurtures Number Sense

In seeking to develop number sense in their students, teachers first must create a classroom environment that encourages student exploration, questioning, verification, and sense-making, and then must select activities that allow students to engage in, and interact with, mathematics. These activities should encourage students to develop their own techniques for finding the answer, to consider many ways to work each problem, to consider the form their answers might take, and to share their reasoning with the class.

In classrooms where number sense is a priority, students are active participants who share their hypotheses, reasoning, and conclusions. Students are given opportunities to create their own procedures for finding solutions. For example, middle school teachers might ask students, "What happens when a positive counting number is multiplied by a fraction less than 1?" Students might explore this question by generating numerous examples using a calculator. As conjectures are made, they can be verified or refuted quickly. Students are given the opportunity to see for themselves that multiplying a positive whole number ($n$) by a fraction less than 1 always results in a product less than $n$. They feel safe discussing their ideas and creating their own procedures for finding answers.

In classrooms where sense-making is a priority, teachers become guides and moderators instead of dispensers of answers. The goal of "one right answer" that is derived from "one preferred algorithm" is replaced by the goal of multiple solution strategies that are generated by, and make sense to, the students. The emphasis shifts from finding the specific solution to investigating how the answer was obtained.

♦ ♦ ♦ ♦ ♦ ♦ ♦ ♦

#### *Use Student Writing*

Writing assignments can be used in a mathematics classroom as an effective method for helping students nurture their sense of number. Students can be asked to write about the results of group activities or to keep journals in which they enter ideas generated by specific tasks. As they write, they may formulate new ideas or questions, which may serve as a reference for further question posing and discussion. Later, as they reread their journals they can see how their conceptions have been changed and expanded by further exploration and discussion. The following excerpt from a fifth-grade student journal illustrates how children might record their thoughts:

> Today I found out that I can't count to 1 000 000 in 1 day. It would take me more than 10 days and I would even have to count all night without stopping. A million is a lot bigger than I thought. I wonder how long it would take to count to a trillion?

To determine students' attitudes and beliefs about large numbers, the teacher may also want to use students' self-assessment data. One way to obtain this type of data is to ask students to write a brief report about counting to a million. Some questions that could be used to guide student writing include the following:

1. How did you feel when you read the problem? Excited? Confused? Frustrated? Why?
2. How do you feel about working on and solving problems that involve large numbers?
3. Did you ever feel that you wanted to give up and not solve the problem? When? Why? What did you do when you had this feeling?
4. Would you rather have worked on this problem by yourself or with others? Why?

Other ideas for using self-assessment data from students can be found in Charles, Lester, and O'Daffer (1987, pp. 23–29).

#### *Use Process Questions and Class Discussion*

The use of process questions (those that require more than a simple factual response) is an important teaching practice for promoting number sense. Consider the following conversation between a teacher and the class:

*Teacher:* I'm thinking of two fractions. Their sum is between 0 and 1. What can you tell me about the fractions?

*Scott:* Both are less than 1.

*Teacher:* Yes, what else?

*Scott:* Are they both less than 1/2?

*Teacher:* Good question, can anyone answer that?

*Kelly:* I can. They don't have to *both* be less than 1/2. Like one can be 3/4 and the other real small, like 1/10.

*Scott:* Oh, I see. But if one is greater than 1/2, the other one must be less than 1/2.

*Teacher:* Good point. Can anyone tell us anything else about my fractions?

*Nicole:* If their sum is less than 1, their product is also less than 1.

*Teacher:* What do the rest of you think about that idea?

The teacher's questions prompted the students to examine various examples and to prove or disprove their initial thoughts. In promoting number sense, it is important for students to see that it's OK to be wrong. What is valued most is reflection on answers so that they can be proved either right or wrong. The teacher in this example assumed the critical role of questioner, which caused students to continue reflective thought about the concept and encouraged other students to be involved in the process.

Teachers must also be flexible and ready to capitalize on comments students make as they explore topics of interest to them. For example, while discussing large and small numbers, one student remarked, "If you wanted to count to a trillion in a lifetime, you would have to be able to count pretty fast!" When the teacher asked the student to explain his reasoning, the student replied: "Well, I know it takes ten days to count to a million because we tried that in class. So, I multiplied ten days times a thousand. That tells me how long it takes to count to a billion (10 000 days). Then I multiplied 10 000 days by another thousand. That tells me it would take 10 million days to count to a trillion." The teacher followed up by asking the student to think about how long 10 million days is and to report the result back to the class.

***Number sense is strengthened when children are encouraged to "poke around" with numbers, gaining insight into the relative effects certain changes have on their solutions. (Whitin 1989, p. 28)***

Manipulative materials or models can also be helpful as vehicles for productive discussions. As they enrich the learning environment, they also give students a means of proving or refuting their conjectures. Providing students with concrete materials may motivate them to initiate new investigations and to continue the same process of validating or rejecting hypotheses that arise from explorations. For example, consider the following activity:

Suppose the circle drawn on this standard piece of paper (8 ½" by 11") is copied on a copy machine that has enlargement and reduction capabilities (fig. 3). Consider these questions:

1. How large will the diameter of the circle be if you copy it at 100 percent? At 50 percent? At 150 percent?
2. Suppose you copy this drawing at 50 percent. Then you decide you want to copy it again to make it look like the original. What size enlargement would you ask the machine to make?
3. Suppose you copy this drawing at 150 percent. Then you decide you want to copy it again to make it look like the original. What size reduction would you ask the machine to make?

Having access to a reduction/enlargement copy machine would provide the necessary reinforcement to validate the proposed answers.

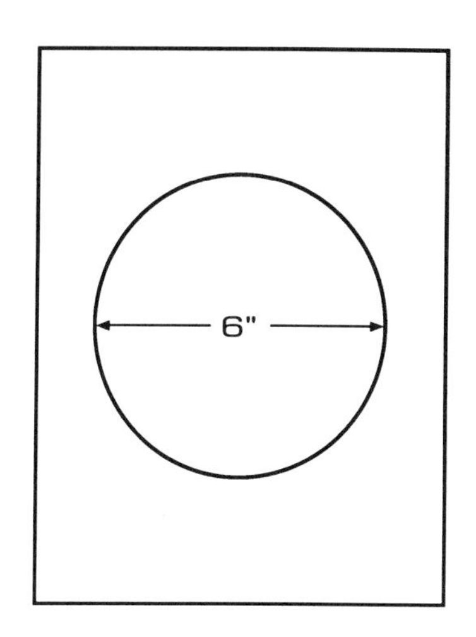


***Fig. 3***

Manipulatives give students the opportunity to experience several representations for different types of numbers or other mathematical concepts. The diagram in figure 4 shows an easy fraction model that every child can make (a template for this model is provided in Appendix 2).

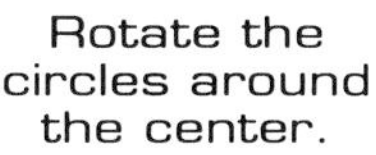


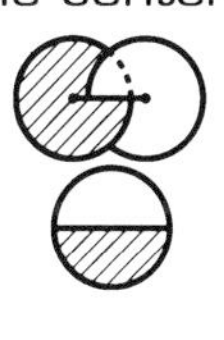

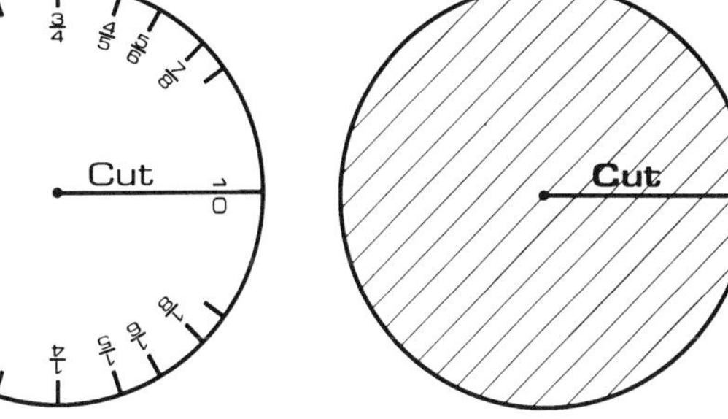


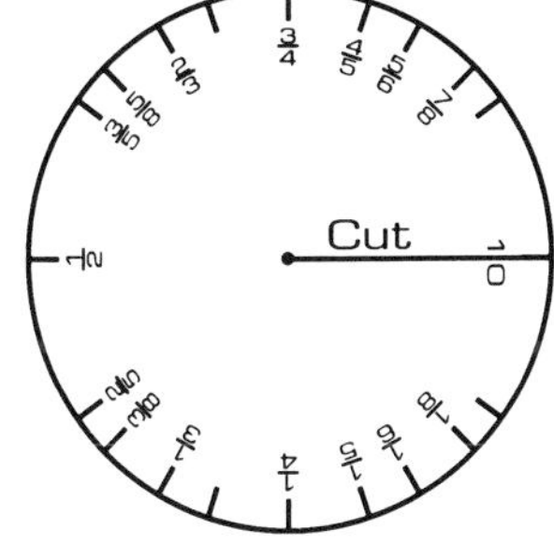

***Fig. 4***

◆ ◆ ◆ ◆ ◆ ◆ ◆ ◆

One side of the circle will have fraction names listed, and the other side can be used by the teacher to pose various questions. For example, the teacher can ask a student to show and name a fraction that is a little less than 1/2; between 1/2 and 3/4; or close to but greater than 1/2. (See Activity 17 for a complete description of this activity.)

The calculator plays an important role in a classroom environment that emphasizes exploration and investigation. By using the calculator to perform tedious computations and test conjectures, students can focus on the process of deriving a solution and on the meaning of the answer once it is computed. Students may be more eager to explore properties of numbers when calculators are accessible. For example, suppose the teacher poses the following question to the class: "What happens when you multiply a positive whole number by a decimal between 0.9 and 1.1? Use your calculator to explore this question; generate your conjecture; and make a list of examples that support your conjecture." In this activity, students are encouraged to generate many examples and to draw a conclusion based on these examples. The calculator makes this exploration efficient and free from distraction.

***Focus on Student-generated Methods of Solution***

***...the educational issue is to instill more widely in the population of students the fact that it [mathematics] should all make sense and that there should be some reasonableness to all this. (Silver, p. 13 in Sowder and Schappelle 1989)***

In promoting number sense, teachers select activities that encourage students to develop their own techniques for finding the answer, to consider many ways to work the problem, to consider the form their answers might take, and to share their reasoning with the class. Teaching for number sense involves a quality over quantity attitude to problem completion. That is, rather than attempting to work as many problems as possible in a given period, students should focus on understanding a given problem by looking at it from multiple perspectives. Strategies offered by students must be justified; when considering each answer, students learn to ask, "Does this make sense?"

Encouraging students to create their own methods for solving problems prepares them to consider traditional methods at a later stage and to view standard algorithms as yet another means of producing sensible answers. They are also more likely to consider variations and further exploration. For example, the following problem was presented to a sixth-grade class:

> Suppose you have saved $63. You find a used video game system that you would like to buy. The seller is asking $180. You earn $10 a week doing odd jobs. How long will it take you to earn enough money to buy the game?

Although adults might use a "standard" solution strategy, such as (180 – 63) ÷ 10, other strategies were generated by students as they solved the problem. For example, one student said that he entered 63 into the calculator (because that was what he had to start with), then added his first week's allowance, then another week's allowance, then another, and so on, until the display showed that he had at least $180 (twelve weeks). Another student said that it would take at least ten weeks (in ten weeks she could earn $100 but that still wouldn't be enough), so in ten weeks she'd have $163 and would need only two more weeks to have enough.

Allowing students the freedom to use strategies that are intuitively obvious to them helps them to feel more comfortable in the problem-solving process. At some stage it also helps them appreciate the efficiency of standard algorithms.

***Focus on Concepts***

Expecting students to develop their own solutions helps them internalize mathematics in a way that makes sense to them. Four fifth-graders were given five candy bars and were told to divide them equally among the group. They wrote the following report:

> Each person got one whole candy bar. One was left over. We cut that candy bar into six pieces and passed out one piece to each. Two pieces were left. We cut those into halves, to make four pieces—one for each. We each had one candy bar plus ⅙ and 1⁄12. Joel said, "Couldn't we just take our last candy bar and divide it in half and half again?" We found out that the ⅙ and 1⁄12 were really ¼. (Burns 1988, p. 31)

These students are developing a deeper understanding of the meaning of fractional remainders as well as of the meaning of adding fractions with unlike denominators.

To help students see the connections between seemingly different calculations, problems such as the following can be presented:

> Use the calculation $3\frac{3}{4} \div \frac{1}{4} = 15$ to determine the answers to these questions:
>
> a. $3\frac{3}{4} \div \frac{1}{2}$ b. $3\frac{3}{4} \div \frac{1}{8}$

Students are encouraged to explore the relationship between dividing by ¼ and dividing by ½ or ⅛. The teacher might guide this exploration by asking, "Which answer (a or b) will be larger than 15? Why? How much larger?"

#### *Accept and Encourage Appropriate Computational Methods*

Number sense can be promoted by ensuring that students learn to calculate in a variety of ways using written, mental, approximate, and electronic methods. Both mental computation and estimation offer vehicles to encourage invention of strategies and alertness to sensible answers.

To offset the often mechanically performed pencil-and-paper calculations seen in school, mental calculation should be included as an accepted method of computation.The methods used to calculate mentally differ from the more standard algorithms used in written arithmetic—they are variable and highly flexible, often depending entirely on the numbers appearing in the calculation and on the personal preference of the calculator. Mental methods generally involve quantities and number relationships (elements of number sense), whereas written algorithms generally involve digits and bookkeeping rules.

Teachers can encourage students to calculate mentally whenever possible, to explore a number of different ways to calculate answers mentally, and to share their techniques with the class. For example, given the problem $5 \times 96$, one student might change the problem to $10 \times 96/2$, another might think $5 \times 8 \times 12$ , and another might use the distributive property and compute $(5 \times 90) + (5 \times 6)$ or $(5 \times 100) - (5 \times 4)$.

Probing questions and concrete analogies can be used to initiate exploring alternative methods of mental calculation. For example, after giving students the problem $25 \times 49$, the teacher might ask, "Can anyone create a problem from this calculation by substituting the word *quarters* for the number 25? Does it help to find the product of these two numbers if we think about the 25 as a quarter? How?"

Approximate computation or estimation is another important tool for encouraging students to use what they already know about numbers to make sense of numerical situations. Oftentimes, this means that students use their own benchmarks to judge the reasonableness of a situation.

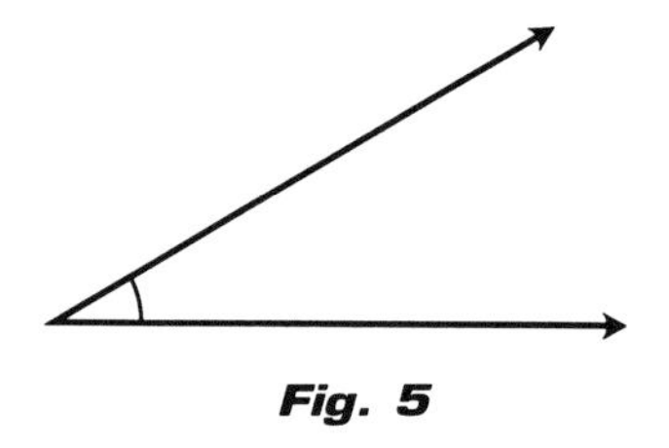

***Fig. 5***

For example, a student using a standard protractor to measure the angle in figure 5 is not likely to read the wrong scale and report 150 degrees as the measure if a 90-degree angle has been established as a referent. In the same way, a student who has been encouraged to estimate fractions near 0, $\frac{1}{2}$, and 1 will understand that the sum of $\frac{2}{5}$ + $\frac{4}{9}$ must be less than 1 since both fractions are less than $\frac{1}{2}$.

Encouraging students to consider these benchmarks or referents is a way of helping them develop better conceptual understanding of fractions, decimals, and percentages. This intuitive understanding is a priority and should precede the study of operating with fractions, decimals, and percentages.

An important role for teachers in developing number sense is helping students learn to ask themselves key questions before, during, and after the solution process. For example, what type of number do I expect for an answer to this problem? About how large will the answer be? What is the biggest or smallest value I expect? After completing a calculation, students should determine whether or not the answer is consistent with what they expected. These steps may help sensitize students to magnitude errors as well as prevent them from checking their answer by repeating the same computational error a second time.

An excellent activity that focuses attention on characteristics of a solution is "No Answer Please" (Glatzer and Glatzer 1989). In this activity, students must list as many facts as possible about an answer but are asked not to give a numerical value. For example, given the problem $\frac{5}{4}$ + $\frac{1}{3}$, students could report that the answer must be greater than 1 because $\frac{5}{4}$ is greater than 1, that the answer is greater than 1 $\frac{1}{3}$, that it is less than 2, and so on. The responses rely on an understanding of fractions rather than on the ability to apply a learned algorithm.

The emphasis on reasonable answers can extend to everyday situations. Students can be encouraged to evaluate the numerical reasonableness of what they read or see on television and to be alert for examples of the misleading use of numbers. For example, a recent television commercial seen in Canada stated that in a survey of football fans, 63 percent of the fans surveyed preferred the taste of one soft drink, whereas 38 percent preferred another. The commercial informed the viewer—in relatively small print—that these results were based on "the number of fans who participated and expressed a preference." This commercial can be used to promote discussion: Is the conclusion accurate? Do we have enough information to determine this? Since we know neither the total sample size nor the breakdown of the results, it is difficult to judge the conclusion. The following possible results demonstrate how uninterpretable the results of the survey are.

| | Number of Fans | | | |
|---|---|---|---|---|
| No preference | 2 | 92 | 992 | 9 992 |
| Prefer soft drink A | 5 | 5 | 5 | 5 |
| Prefer soft drink B | 3 | 3 | 3 | 3 |

For example, 1000 people might have been surveyed, with 5 preferring soft drink A, 3 preferring soft drink B, and 992 indicating no preference.

***Number sense develops over time. This development is best nurtured if the focus is consistent, day by day, and occurs frequently within each mathematics lesson. (Thornton and Tucker 1989, p. 21)***

#### *Provide Process-oriented Activities*

By establishing a classroom atmosphere that encourages exploration, discussion, and thinking and by selecting appropriate problems and activities, the teacher can cultivate number sense during all mathematical experiences. Activities that promote good number sense by concentrating on process have several common characteristics. They encourage students to think about what they are doing and to share their thoughts with others. They promote creativity and investigation and allow for many answers and solution strategies. They help students know when it is appropriate to estimate or to produce an exact answer and when it is appropriate to compute mentally, on paper, or with a calculator. They help students see the regularity of mathematics and the connection between mathematics and the real world. Process-oriented activities also convey the idea of mathematics as an exciting, dynamic discovery of ideas and relationships. Examples of process-oriented activities have been highlighted throughout this discussion. More are provided beginning on page 16.

***...number sense is not a discrete quality that a person either has or does not have; rather it is a continuous quality that is possessed at different levels, and these levels have the potential for constantly expanding to reflect new experiences and higher levels of understanding. (Reys, p. 65 in Sowder and Schappelle 1989)***

### *EVALUATING NUMBER SENSE*

Evaluating number sense is an ongoing process that involves listening to students' explanations and observing the connections students make as their base of understanding builds. This evaluation process is much more qualitative than quantitative and cannot be easily measured by paper-and-pencil tests. Rather, evaluating number sense is best accomplished through class discussion, dialogues with individual and small groups of students, and examination of individual journal entries. For example, consider this classroom episode (see Activity 24 for a complete description):

*Teacher:* Class, look at the number line I've presented. I'm going to ask you to use it to answer a few questions. First of all, name the number that might be represented by the letter *B*.

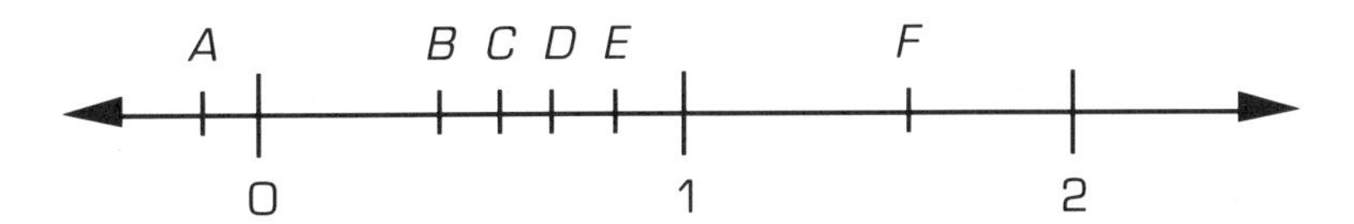


*Lauren:* It looks a little less than ½ so I'll say ⅜.

*Scott:* I think that's a little too small, it's closer to ½.

*Teacher:* Scott, what makes you say ⅜ is too small?

*Scott:* Well, ⅜ is 0.375, which is near ⅓, and I think that *B* is closer to ½ than ⅓.

*Teacher:* OK, Lauren, what do you think about Scott's suggestion?

*Lauren:* He's probably right. Three-eighths is a ballpark figure. If I had to give a closer guess, I'd say that *B* is about at the point 0.45.

*Abe:* Is it OK to use decimals?

*Teacher:* What do the rest of you think? Can we represent *B* using either decimals or fractions?

*Christy:* Sure, why not? They're both ways to express numbers that aren't whole numbers.

*Teacher:* OK. Let me give you another question to consider. Suppose I multiplied the number represented by *E* by 26. What answer would you expect?

*Whitney:* I don't think we have enough information. How can we multiply a number by a letter?

*Teacher:* Think about this, Whitney. What do you know about the number represented by the point $E$?

*Whitney:* Well, I know it's smaller than 1. But I don't know what it is.

*Teacher:* True. What else do you know about $E$?

*Whitney:* I guess it's bigger than $\frac{1}{2}$ because it's closer to 1 than to 0.

*Teacher:* Good. Now think, what will happen if I multiply 26 by a number between $\frac{1}{2}$ and 1?

*Whitney:* I'm not sure.

*Teacher:* OK, can anyone help Whitney?

*Rustin:* If we multiply 26 by $\frac{1}{2}$ we get 13, and if we multiply 26 by 1 we get 26, so the answer has to be between13 and 26.

*Whitney:* I get it. And since $E$ is closer to 1 than $\frac{1}{2}$, the answer will probably be closer to 26 than 13.

*John:* Right. I'd say it's probably 22.

*Teacher:* Can you tell us why you picked 22?

*John:* Not really, it just seems right.

*Terry:* I know a way to justify it. If you think about $E$ being 0.9, then $0.9 \times 26$ is about the same as $0.9 \times 25$, or 22.5.

*Teacher:* Good thinking on both John's and Terry's parts. Let me give you a new question. Suppose....

This classroom episode demonstrates a rich discussion session that challenges students to think and to justify their thinking. It also allows students at various levels to contribute. The teacher gets a good indication of the students' levels of number sense by their contributions and their ability to think in multiple ways about problems. As a result of this classroom conversation the teacher may take a variety of actions: make notes in individual student portfolios; make plans to talk with Abe individually to explore more fully his concept of fraction/decimal relationships; make a note in tomorrow's lesson plan to pose a similar problem; develop a writing assignment to be used later in the week to gather expanded individual documentation of the concept; or, use a checklist as an ongoing assessment record of students' attitudes and beliefs during class discussion (see figure 6).

| | Behavior | | | | |
|---|---|---|---|---|---|
| | Shows willingness to consider alternative ideas | Demonstrates self-confidence | Exhibits tolerance for ambiguity | Shows eagerness to think for self | Demonstrates an appreciation of need for logical thinking |
| Lauren | ✓ | | | | |
| Scott | | | | | |
| Abe | | | | | |
| Christy | | ✓ | | ✓ | |
| Whitney | | | ✓ | | |
| Rustin | | | | | |
| John | | | | | |
| Terry | | | | | ✓ |

***Fig. 6. Sample checklist to assess attitudes and beliefs during classroom discussion.***

♦ ♦ ♦ ♦ ♦ ♦ ♦ ♦

There are various ways to use a checklist like the one presented in figure 6. First, teachers who are interested in assessing classroom climate and students' attitudes and beliefs should determine what aspects of student behavior are important to be recorded. The behaviors can be listed in the column headings. The column headings provided in figure 6 are used to illustrate some of the behaviors related to number sense that might be appropriate to observe.

Second, recording information in the checklist could be done by using checks (✓) or by using "+" and "–" symbols to indicate the presence or absence of a specific behavior. The teacher's interpretation of individual students' behaviors may cause supporting entries in the checklist to vary. For example, Lauren's reaction to Scott's statement may support an entry in the first column because she was willing to consider Scott's comment and she may believe that there is not just one way to do mathematics. Christy's statement reflects self-confidence (although teachers may want to have more evidence before making an entry in the second column). Furthermore, Christy shows an eagerness to think for herself. At first, Whitney expressed doubt when she was told to multiply 26 by the number represented by *E*, but then she demonstrated a willingness to resolve the ambiguity. Terry's statement supports her appreciation of a need for logical thinking.

Presenting problems that challenge students to think logically and to draw on their knowledge is perhaps the best way to evaluate the development of number sense. It requires frequent opportunities for discussion and a selection of rich problems to promote this discussion. The discussion not only allows the teacher to evaluate developing number sense but also provides an opportunity for students to learn from each other. For a more detailed discussion of evaluation, see Appendix 1.

***Number sense builds on students' natural insights and convinces them that mathematics makes sense, that it is not just a collection of rules to be applied. (Howden 1989, p. 7)***

***SUMMARY***

The NCTM *Curriculum and Evaluation Standards* recommends a number of changes in the middle grades mathematics curriculum. Many of these changes are related to the development of number sense. Figure 7 highlights the areas that are to receive increased and decreased emphasis.

| Increased Attention | Decreased Attention |
|---|---|
| Developing number sense | Memorizing rules and algorithms |
| Developing operation sense | Practicing tedious paper-and-pencil computations |
| Creating algorithms and procedures | Finding exact forms of answers |
| Using estimation both in solving problems and in checking the reasonableness of results | Memorizing procedures, such as cross-multiplication, without understanding |
| Exploring relationships among representations of, and operations on, whole numbers, fractions, decimals, integers, and rational numbers | Practicing rounding numbers out of context |
| Developing an understanding of ratio, proportion, and percent | Developing skills out of context |
| | Learning isolated topics |

***Fig. 7. Changes in grades 5–8 mathematics content and emphasis related to number sense as suggested by the* Curriculum and Evaluation Standards.**

Clearly, the changes envisioned call for more emphasis on understanding and conceptual development and for far less attention to skill development.

The role of the teacher in developing number sense is to foster the idea that mathematics learning is a sense-making activity. This is done by creating an environment that encourages students to question, explore, and explain their thinking by motivating them to extend their ideas. The use of manipulatives and mathematical models, discussion sessions, student-generated strategies, and process-oriented activities helps to create the classroom environment that has number sense as a high priority.

### *NUMBER SENSE ACTIVITIES*

Included here is a selection of problems and activities, appropriate for the middle grades classroom, for which the underlying theme is the development of number sense. These activities can be used in varied ways to generate discussion and to extend student thinking about number-related concepts. They might be used to introduce or extend a topic or to close a lesson. The discussion that arises as students describe their thinking will certainly give insight into their thinking and will help in evaluating students' development of number sense.

The ideas presented in the activities are directed to the teacher and are not prepared as blackline masters for immediate student use. The format includes a brief discussion outlining the purpose of the activity. None of the activities is meant to be a complete lesson plan. Rather, they are ideas that can supplement teaching. It is hoped that these problems and activities will serve as springboards for generating additional ideas to help foster the development of number sense. In that spirit, one expanded activity is offered here to illustrate the approach we are recommending: using activities that promote sense-making and generate rich discussion among small groups of students and between students and teacher.

#### *An Expanded Activity*

This activity focuses on developing a sense of fractions between 0 and 1. Students work in small groups as they complete tasks by conjecturing, defending, and justifying their thinking to fellow groupmates. In advance of the activity, prepare sets of fraction cards, one per group of students. Each set should contain nine cards—one each for $1/4$, $9/10$, $2/19$, $4/7$, $7/9$, $1/3$, $12/15$, $2/5$, and $5/8$. Put students into small groups (three to five students per group) and give each group one set of fraction cards.

Begin a whole-class discussion by asking questions to familiarize the class with the fractions they will be working with. Ask questions such as these: What is the smallest fraction in the set? What is the largest fraction in the set? As individual students provide answers to these questions, encourage them to justify their thinking. For example, a student might name $1/4$ as the smallest fraction in the set, completely dismissing the fraction $2/19$ because of the large denominator. Be sure to ask students how they think about $2/19$. Since $2/19$ is a bit of an "odd" fraction, it is important to relate it to a known benchmark. For example, one student might think about $2/19$ in this way: "It's pretty close to $2/20$, which is $1/10$." You might follow this comment with these questions: "Good, that makes sense. Do you know whether $2/19$ is smaller than or larger than $2/20$? How do you think about that?"

The first task of each group is to sort the fraction cards into two piles—one containing fractions less than $1/2$ and one containing fractions greater than $\frac{1}{2}$. Once again, students must justify the placement of each fraction to their groupmates. Be alert to the kinds of explanations students give for placing fractions such as $\frac{4}{7}$ or $\frac{2}{5}$. For example, are students using $\frac{1}{2}$ as a benchmark from which to judge these fractions?

The second group task is to order the nine fractions from smallest to largest without using paper and pencil. Instead, encourage students to use benchmarks such as 0, $\frac{1}{2}$, 1, and other common fractions to place each card. Again, encourage the students to discuss their thinking within their groups. When all groups have finished, ask students to tell which fractions were most difficult to place. Which were easiest? Why?

The final group task is to pair the fraction cards so that the fractions in each pair add to about 1. There will be four pairs of cards and one "leftover." For example, students might put $\frac{2}{5}$ and $\frac{4}{7}$ together, explaining that because one is a little smaller than a half and the other is a little bigger than a half, their sum is close to a whole. Follow-up questions could include these: Which pairs give a sum larger than a whole? Which give a sum smaller than a whole? Which sums are closest to a whole? What strategies were used in pairing up the fractions? For example, one group of students explained their strategy by saying that they used the ordering of fractions from the earlier task and simply picked up the smallest and the largest fraction to form their first pair, then picked up the next smallest and next largest to form their second pair, and so on. Other groups used different strategies, including one group that mentally converted each fraction to its decimal equivalent, then selected the "best" pairs.

An additional follow-up idea might be to ask each group to hold up its "unused" fraction for the last group task. Do all groups hold up the same fraction? Should they?

This activity can be varied in a number of ways, including changing the numbers that appear on the fraction cards. For example, the teacher might include fractions that are more or less common, fractions that are closer to 0 or to 1, and fractions that contain "ugly" denominators—large or prime numbers. The important ingredient of this activity is the focus on making sense of each fraction by transforming it in some way that makes sense to the individual. For some students, that may mean "rounding" to a common fraction. For other students, it may mean converting to an approximate decimal form. For still other students, it may mean relating each fraction to some benchmark or using a mental model to represent the fraction as part of a whole. It is important that all students see that there are several ways to think about a fraction. This can be accomplished by having students exchange ideas and work together on tasks that challenge them and cause them to think in nonalgorithmic ways.

The remainder of this book suggests some ideas for the reader to try. This set of activities is offered not as a "number sense curriculum" but as a collection of ideas related to number sense and drawn from a variety of topics within the middle grades curriculum. It is our hope that these activities may further illustrate what number sense is and highlight some ways to encourage its development.

◆ ◆ ◆ ◆ ◆ ◆ ◆ ◆

## ***ACTIVITY 1: ESTABLISHING BENCHMARKS FOR WHOLE NUMBERS***

**To the Teacher:** Adults often use benchmarks or common referents to process numerical information. For example, knowing the population of your town might help you judge the size of a crowd attending a concert. (For example, if the high school stadium holds 1 000 people and the report says that 150 000 people attended a rock concert, you might think of the size of the concert crowd as being the stadium filled 150 times.) This activity is designed to see how many students are aware of the size of some commonly used referents. Encourage students to formulate additional questions and problems for studying the implications of these statistics. As an extension, you might want to review with your class the book *In One Day* by Tom Parker (1984), which contains a number of interesting numerical facts about what Americans do in one day.

***Discuss resources available for finding these answers: almanacs, members of Congress, government printing offices.***

***The 1989 world population was 5.192 billion; the 1989 U.S. population was 247.1 million.***

Pose the following questions to students. Allow them first to estimate and then to do research to determine the value that answers each question.

1. Population of the world: ___________

2. Population of the United States: ___________

3. Population of your town: ___________

4. Population of your school: ___________

5. United States government budget: ___________

6. The number of 13-year-olds alive today: ___________

7. The number of graduating high school seniors last year: ___________

8. The number of tons of garbage generated every day: ___________

## ACTIVITY 2:
## USING CONTEXT TO DETERMINE REASONABLE VALUES

**To the Teacher:** Encourage students to work in small groups as they "edit" this paragraph and insert sensible values for those that are missing. As an extension, ask students to find a newspaper story with numerical data. One group of students blanks out all numerical values and the other group must replace these values.

Ask students to help "write" this newspaper story. Review each place where a number is missing. Select from the list below the numbers that will make the story make sense.

PTA Meeting Attracts _________

An audience of about _________ attended the fall Chorus Sing for the PTA on Monday, October _________. The parents, teachers, and friends heard _________ songs and saw _________ musical skits. After the entertainment, last year's school-attendance award was presented to Sally Singer. She attended all _________ days of school last year. It was her _________ year of perfect attendance. Then the parents and children had snacks. _________ cupcakes were eaten and _________ glasses of punch were drunk.

***Evaluation idea: The short-answer format of this activity could be used to analyze students' understanding of number to assist in planning for instruction.***

Numbers to use:

180 hundreds fourth 24th 400 six 526 616 seventeen

Source: Fennell, Francis, Larry Houser, Donna McPartland, and Sandra Parker. "IDEAS." *Arithmetic Teacher* 31 (September 1983): 27–32.

## *ACTIVITY 3:*
## *VERIFYING MATHEMATICAL STATEMENTS*

**To the Teacher:** This activity promotes consideration of the reasonableness and significance of published statistics. It also reinforces the practice of questioning and verifying mathematical data. You may want to continue this activity with additional facts. Several excellent resources for "facts" include *In One Day* (Parker 1984), *The World Almanac and Book of Facts* (1990), and *Mathematics and Global Survival* (Schwartz 1989).

Each student or group of students is given a fact obtained from a published source. Students are asked to verify the accuracy of the statement that follows the fact, using whatever resource material or calculations they need. Each individual or group prepares a report of their investigation for the class.

***Ask small groups of students to make a plan to verify the statement. What information is needed? For example, for the first "fact," what is the distance from New York City to Hollywood?***

FACT: In one day Americans buy 50 000 new television sets (60 percent of them are color sets).

> If you spread the 50 000 television sets evenly along the road between New York City and Hollywood, they would be just over 300 feet apart—less than a minute's walk. Television addicts could easily walk from one TV to the next during commercials and never miss the show.

FACT: In one day Americans driving on unpaved roads stir up 81 000 tons of dust.

> That's enough dust to cover a football field—the Dust Bowl—to a depth of forty-eight feet.

***You might want to use this activity as homework, involving parents in the task.***

FACT: In one day trains in this country haul 4 million tons of material in 58 000 railroad cars.

> That's equivalent in weight to picking up every person west of the Mississippi, except the Californians, and moving them back east. The 58 000 cars, made into one long train, would stretch 600 miles.

Source for facts: Parker, Tom. *In One Day.* Boston: Houghton Mifflin, 1984.

♦ ♦ ♦ ♦ ♦ ♦ ♦ ♦

## ***ACTIVITY 4:***
## ***REASONABLE VALUES IN STATEMENTS***

**To the Teacher:** This activity highlights the need to be sensitive to the reasonableness of facts we see around us (newspapers, television, and so on). Encourage children to generate or bring to class other pieces of "nonsense news."

Display each news clipping and ask students to evaluate it. Does it make sense? If not, why?

NINTENDO HOTLINE: Five hundred Nintendo players were asked to list their favorite games. The results are shown in this graph:

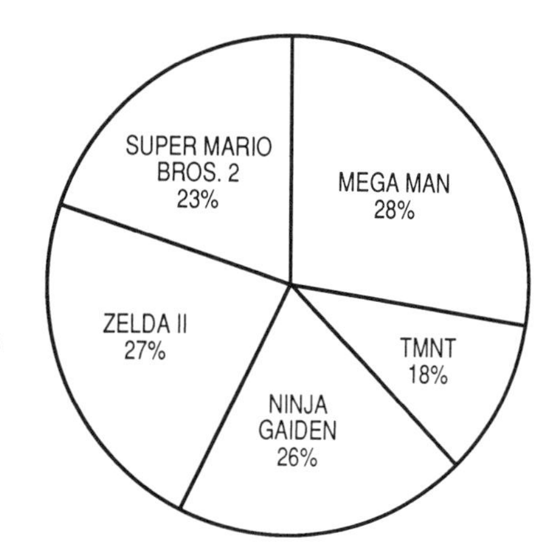


ADVERTISEMENT: Are you looking for a quality preschool? At Eager Beavers Learning Academy, we maintain a 1:10 teacher-to-student ratio, keeping everyone happy. Just look at the smiling faces in the photo of our three well-trained teachers and our fifty full-time students.

***Encourage students to use a calculator as they investigate the Mars space launch.***

SPACE LAUNCH TO MARS: NASA says all systems are "go" for the Wednesday launching of spaceship Phobos on its voyage to Mars. Traveling at 8000 miles per hour, the ship is expected to make the 34-million-mile trip in about four days.

INCOME TAX TIME: The IRS reports that the new income tax forms are easier than those used last year. "Most people should be able to complete the forms by themselves this year," said Director Tim Fordy. "We expect 23 percent of taxpayers will be able to complete their forms without needing professional assistance."

## ACTIVITY 5:
## REASONABLE ANSWERS

**To the Teacher:** Ask students to clear their desks, then pose the following problem.

***After posing this question, encourage students to rely on their own common sense to think about what a reasonable answer would be.***

Problem: Suppose Ms. Jones, an experienced painter, can paint a wall in three hours, whereas a rookie painter, Mr. King, paints the same wall in seven hours. If the two painters work together to paint the wall, how long will it take them to complete the job? Choose the most sensible answer. Explain your decision.

a. 21 hours

b. 10 hours

c. between 5 and 10 hours

d. 5 hours

e. 4 hours

f. 3 hours

g. 2 hours

h. ½ hour

---

## ACTIVITY 6:
## EXACT AND APPROXIMATE NUMBERS

***As a follow-up activity, ask students to think about situations when exact values are needed and other situations when approximate values are necessary and useful.***

**To the Teacher:** Ask students to investigate the use of approximate and exact numbers in newspaper articles. Use this activity to help students appreciate the everyday use of estimates.

Give students copies of the front page of various newspapers. Ask them to use a marker to circle numbers used in headlines and articles. Next, have students review the context for the use of each circled number to determine if it is an exact or an approximate value. For example, do the numbers in the headlines below refer to exact or approximate values?

U.S. population tops 220 million

Lottery winner earns over $200 000 annually

Stocks fall 5.4%

## *ACTIVITY 7: HOW BIG IS A MILLION?*

**To the Teacher:** Manipulatives such as beans are excellent for teaching place value and the relative size of numbers, but they can become cumbersome with numbers larger than 100. This activity was designed and conducted by a sixth-grade teacher and his class to help the students better understand the size of one million.

A class discussion led to the idea that the students could better understand the size of one million if they actually collected a million pieces of some object. The discussion led to a brainstorming session about what kinds of objects would be reasonable to collect. A final decision was made to collect toothpicks because of their size and availability. On a bulletin board, the class made a large place-value chart where they recorded the current toothpick count. Each day the students brought in toothpicks, which were added to the collection as the chart was updated.

The project lasted 1½ years. The students not only better understood place value, they also had a better sense of the size of one million. It took twelve 39-gallon lawn and leaf bags to haul the toothpicks away.

Extension Activities:

1. Estimate—
   a. the length the toothpicks would stretch if laid end to end;
   b. the weight of the one million toothpicks;
   c. the cost of the toothpicks.
2. Brainstorm about possible uses for the million toothpicks collected (e.g., art projects, science projects, construction of geometric figures).

***To evaluate the level of number sense, have students respond to one or more of the extension questions in journal format.***

Source: David Hoffman, Captain Elementary School, Clayton, Missouri.

## *ACTIVITY 8:*
## *EXPLORING THE SIZE OF A MILLION DOLLARS*

**To the Teacher:** This activity explores whether one million dollars will fit into a standard suitcase. If so, how large would the suitcase need to be? How heavy would it be? You may have students work in small groups to explore these questions.

Materials: One thousand or more fake dollar bills (play money or rectangular sheets of paper the approximate size of a dollar bill)

Begin the investigation by telling the following story:

***Calculators should be available to facilitate and expedite the computation for analysis.***

> Just as you decide to go to bed one night, the phone rings and a friend offers you a chance to be a millionaire. He tells you he won $2 million in a contest. The money was sent to him in two suitcases, each containing $1 million in one-dollar bills. He will give you one suitcase of money if your mom or dad will drive him to the airport to pick it up. Could your friend be telling you the truth? Can he make you a millionaire?

***The dimensions of a one-dollar bill are approximately 6 inches by 2 1/2 inches. Twenty one-dollar bills weigh approximately 7/10 of an ounce.***

Involve students in formulating and exploring questions to investigate the truth of this claim. For example—

1. Can $1 000 000 in one-dollar bills fit in a standard-sized suitcase? If not, what is the smallest denomination of bills you could use to fit the money in a suitcase?
2. Could you lift the suitcase if it contained $1 000 000 in one-dollar bills? Estimate its weight.

### *ACTIVITY 9:*
### *COUNTING TO A MILLION*

**To the Teacher:** This activity gives students the opportunity to think about how long it would take to count to a million. The activity may be extended to have students determine how long it might take to count to a billion or a trillion.

Materials: Computer

Depending on students' programming experience, have them write a counting program to explore the time it takes for a computer to count to one million. For example, in BASIC the program might look like this:

```
10 X = X + 1
20 PRINT X
30 GOTO 10
```

***Using an Apple IIe, this program counts to 2000 in one minute. If kept running, it would count to one million in about 500 minutes, or over eight hours!***

Ask a student to type the program and begin to run it. Encourage students to watch the monitor as the computer counts. Let it run for a few seconds, and then break into the program (i.e., "control C" aborts the program).

Ask students to make guesses, based on their observation, about how long it will take for the computer to count to a million. (Note: Guesses generally range from ten minutes to an hour.)

Now time the computer as it counts to 100 000. Use this information to answer these questions: How long will it take to count to a million? A billion? A trillion?

Source: David Hoffman, Captain Elementary School, Clayton, Missouri.

♦ ♦ ♦ ♦ ♦ ♦ ♦ ♦

## *ACTIVITY 10: EXPLORING THE SIZE OF ONE-MILLIONTH*

**To the Teacher:** This activity provides students an opportunity to explore the decimal number 0.000 001 (one-millionth).

***This is a good activity for small groups of students to explore.***

Using an eye dropper, show students one drop of water. Tell them to imagine that this drop of water represents one millionth of something. Their task is to determine what the "something" is; if the drop of water is one-millionth, what is one?

Help the students understand that one million drops of water will represent the unit, or "something," we are talking about. They will need to explore ways of estimating the size of one million drops of water.

Throughout this activity, encourage students to use their understanding of place value to form their estimate. For example, they might count 100 drops of water (approximately 1.5 teaspoons) and record:

If 1 drop = 0.000 001, then 100 drops = 0.000 1.

Next, students might duplicate the 100 drops ten times (this makes about 100 ml of water) and record:

If 100 drops = 0.000 1, then 1000 drops = 0.001.

Continuing this measuring, counting, and recording process, the students will arrive at an answer: One drop represents one-millionth of the amount of water used to take a bath (between ⅓ and ½ of a bathtub full of water)!

---

## *ACTIVITY 11: EXPLORING THE SIZE OF ADDENDS*

***Encourage students to verbalize their thinking. For example, one student responded this way to question 2: "It can't be. If four of the numbers were as small as they could be—each 10—that would leave 60 for the fifth number. There's no way to make one of the numbers larger than 60, unless you can use single-digit numbers."***

***Evaluation idea: Observing and interacting with students during this activity can provide a means of making decisions regarding future instruction.***

**To the Teacher:** You can use this problem as an opening activity to promote discussion.

The sum of five two-digit numbers is less than 100. For each of the following statements, decide whether it is necessarily true, necessarily false, or possibly true. Explain your reasoning.

1. Each number is less than 20.
2. One number is greater than 60.
3. Four numbers are greater than 20, and one is less than 20.
4. If two numbers are less than 20, at least one is greater than 20.
5. If all five numbers are different, then their sum is greater than or equal to 60.

Source: Markovits, Zvia, Rina Hershkowitz, and Maxim Bruckheimer. "Research into Practice: Number Sense and Nonsense." *Arithmetic Teacher* 36 (February 1989): 53–55.

♦ ♦ ♦ ♦ ♦ ♦ ♦ ♦

## ACTIVITY 12: EXPLORING THE SIZE OF FACTORS

**To the Teacher:** Use this problem to promote discussion about place value and multiplication. Encourage students to generalize their solution and apply it to similar problems.

Ask students to copy the following diagram on their paper:

```
 ___ ___ ___
×    ___ ___
 ___________
```

***Encourage the use of the calculator to explore various possible solutions.***

Ask them to choose any five digits (e.g., 1, 2, 3, 4, and 5) and to consider each of these questions:

***There are sixty different possible products. A table could be developed to demonstrate them.***

How would you arrange the digits in the multiplication problem to produce the largest product possible?

How would you arrange the digits in the multiplication problem to produce the smallest product possible?

How many different products are possible using the five digits selected?

Next, have the students choose five different digits and answer the questions above.

Suggest a method for arranging any five digits as a three-digit-by-two-digit multiplication problem to form the largest and smallest possible product.

### *ACTIVITY 13:*
### *COMPARING NUMERICAL EXPRESSIONS*

**To the Teacher:** Use this activity to stimulate discussion regarding the effect of operating on whole numbers. Students should be able to complete each equation without doing any computation but rather by carefully considering the numbers and operations involved in each sentence. In fact, students who perform each computation will likely take more time than those who think through each problem.

Ask students to compare each side of the expressions listed and, without calculating, to decide if <, = , or > completes each sentence. Ask students to justify their thinking.

1. $373 + 574 + 936 \ \square\ 456 + 608 + 988$
2. $723 + 439 + 601 \ \square\ 723 + 439 + 603$
3. $420 + 420 + 420 + 420 \ \square\ 5 \times 420$
4. $72 + 73 + 74 \ \square\ 73 \times 3$
5. $30 \times 13 \ \square\ 31 \times 13$
6. $350 \div 8 \ \square\ 350 \div 10$
7. $1348 \div 56 \ \square\ 1328 \div 56$
8. $15 \times 16 \times 2 \ \square\ 2 \times 16 \times 15$
9. $543 + 709 + 146 + 32 \ \square\ 709 + 32 + 543 + 146$

***One student explained her thinking for question 4 in this way: "They are the same, because if I take one from the 74 and give it to the 72, I'd have all 73s."***

## ACTIVITY 14:
## ESTIMATING WHOLE NUMBER PRODUCTS

**To the Teacher:** In this activity, students choose factors to produce products adjacent to each other in the grid. One calculator is needed after the factors are chosen. Encourage students to make their estimates using strategies such as front-end or rounding while using the product of the units digits as an additional clue.

***Generate additional games by choosing a new assortment of factors (e.g., all decimal factors) and creating an appropriate grid of the possible products.***

Four-in-a-Row

Choose one member of the class to be the "calculator." Provide that student with a calculator. Divide the rest of the class into two teams. To begin, display the grid and the factor board below. In turn, each team chooses two factors from the factor board. If the product of those numbers is displayed on the grid, the team captures that cell. The first team to capture four cells in a row (vertically, horizontally, or diagonally) is the winning team.

Grid

| | | | | |
|---|---|---|---|---|
| 187 | 1189 | 1769 | 943 | 697 |
| 1403 | 319 | 1219 | 1037 | 437 |
| 901 | 1159 | 323 | 551 | 2501 |
| 1007 | 253 | 1537 | 671 | 391 |
| 583 | 779 | 3233 | 667 | 451 |

Factor board

| | |
|---|---|
| 11 | 29 |
| 17 | 41 |
| 19 | 53 |
| 23 | 61 |

## ACTIVITY 15: ESTIMATING THE SIZE OF AN ANSWER

**To the Teacher:** The idea illustrated in this activity might be used before assigning a group of computation problems. Choose several problems from those to be assigned and ask students to think about how large the answer will be. Ask questions such as these: Will the answer be a two-digit number? A three-digit number?

How Many Digits?

Provide problems such as those listed here. After examining each number represented in the problem, ask students to identify how many digits the answer will contain. Ask them to explain their thinking.

***Encourage students to verbalize their thinking. For example, on question 1, a student might think: "I know the answer will have three digits because it can't be as large as 200 + 700, or 900." This type of discussion provides information to the teacher regarding students' global understanding of each operation and communicates to the students the value of estimation.***

1. 134 + 689
2. 134 + 989
3. 1246 − 348
4. 2054 − 128
5. 12 × 234
6. 5 × 689
7. 2344 ÷ 4
8. 2338 ÷ 14

Source: Reys, Robert E. "Estimation." *Arithmetic Teacher* 32 (February 1985): 37–41.

---

## ACTIVITY 16: FRACTIONS BETWEEN 0 AND 1

**To the Teacher:** This activity helps clarify students' understanding of fractions between 0 and 1. It can be used as a quick warm-up for the mathematics lesson. Several questions provided here have more than one correct answer.

Name That Number!

Ask students questions such as these:

1. Name a fraction between $\frac{1}{2}$ and 1.
2. Name a fraction between $\frac{1}{4}$ and $\frac{3}{4}$, other than $\frac{1}{2}$.
3. Name a fraction between $\frac{1}{4}$ and $\frac{1}{2}$ whose denominator is 10.
4. Name a fraction between $\frac{7}{8}$ and 1. How many can you name?
5. Name a fraction between 0 and $\frac{1}{10}$ whose numerator is not 1.

## ACTIVITY 17:
## ESTIMATING FRACTIONS BETWEEN 0 AND 1

**To the Teacher:** The model suggested here is easy to make and will help you evaluate your students' understanding of fractions between 0 and 1. Encourage students to make estimates using familiar benchmarks (e.g., $\frac{1}{4}$, $\frac{1}{2}$, $\frac{3}{4}$).

Copy the circles illustrated onto light-colored construction paper. Give each student a copy and ask them to cut out the circles and make a cut in the radius of each.

***A template for this fraction model appears in Appendix 2.***

Begin by asking students to put the circles together so that they can see the fractions printed on one side of one circle. Ask questions such as these:

Rotate the circles around the center.

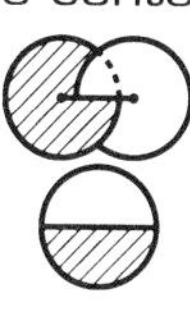

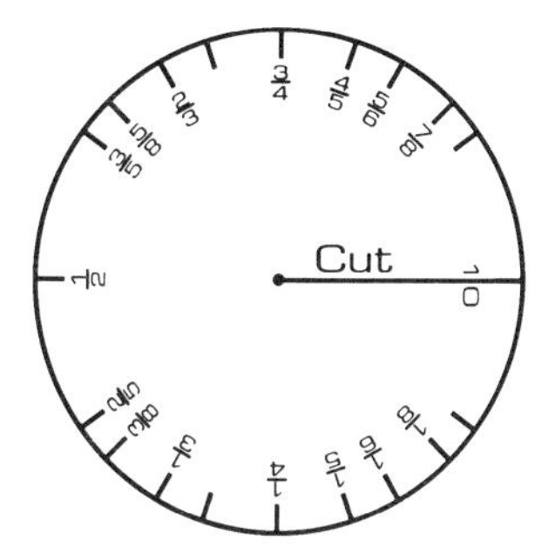


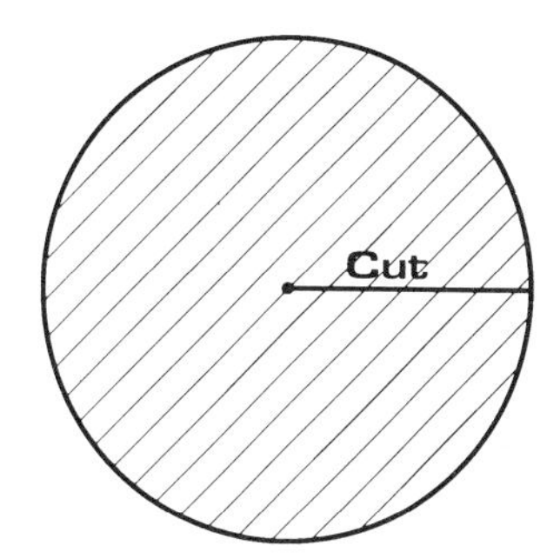


Show $\frac{1}{2}$ of the shaded circle.

Show $\frac{1}{4}$ of the shaded circle. What part of the circle is not shaded?

Show $\frac{9}{10}$ of the shaded circle.

Show a small part of the shaded circle (less than $\frac{1}{4}$). Can you name the part represented?

Ask students to reverse the circle with the printed fractions so that they cannot see the fractions. Ask students if they can—

Show a fraction that is a little bigger than $\frac{1}{2}$. What name can you give it?

Show a fraction that is between $\frac{1}{2}$ and $\frac{3}{4}$. What name can you give it?

Continue asking questions that allow students to show their understanding of the fractions represented.

***Other fraction models should also be used to evaluate students' understanding of fractions (e.g., see Activity 19).***

Source: Hoffer, Alan R., ed. *Mathematics Resource Project: Number Sense and Arithmetic Skills.* Palo Alto, Calif.: Creative Publications, 1978. Used with permission of the University of Oregon.

## ACTIVITY 18:
## IDENTIFYING FRACTIONS NEAR 0, ½, AND 1

**To the Teacher:** This activity helps students think about fractions near the important benchmarks of 0, ½, and 1. You may want students to work in groups to complete each task so that they can share materials.

Materials: Using a different color for each circle, make a set of fraction pieces that represent halves, thirds, fourths, fifths, sixths, eighths, and tenths for each group of students.

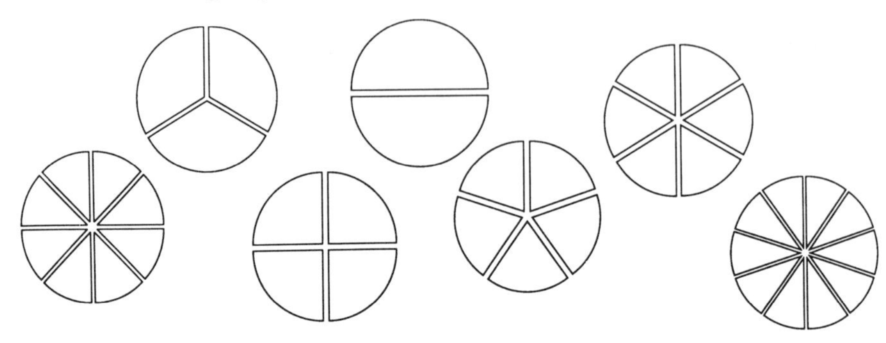

1. Students use one group of fraction pieces (e.g., eighths) at a time to represent fractions near 0, ½, and 1. Record the answers in a chart:

| Fractions near 0 | Fractions near ½ | Fractions near 1 |
|---|---|---|
| $\frac{1}{8}$ | $\frac{3}{8}, \frac{4}{8}, \frac{5}{8}$ | $\frac{7}{8}$ |

Continue with other sets of fraction pieces.

2. After completing the chart, ask students to look for patterns within the column that describes the group. What's alike about all fractions close to 1?

3. Ask students to sort these fractions into three groups (close to 0, close to ½, close to 1) using the generalizations made earlier.

$$\frac{4}{7} \quad \frac{1}{7} \quad \frac{8}{9} \quad \frac{4}{9} \quad \frac{3}{7}$$

$$\frac{5}{12} \quad \frac{2}{12} \quad \frac{9}{11} \quad \frac{6}{11}$$

4. Look at the fractions listed in the column "Fractions near ½." Sort these into three groups (using the models if necessary): those less than ½, those equal to ½, and those greater than ½.

Source: Bezuk, Nadine. "From the File: Easy Pieces." *Arithmetic Teacher* 36 (February 1989): 3.

## ACTIVITY 19:
## VISUALIZING FRACTIONS

**To the Teacher:** Use this activity to evaluate students' understanding of fractions between 0 and 1.

Display the fraction strips using an overhead transparency. Ask questions such as these:

One of the strips shows the fraction $3/8$. Which one is it?

One of the strips shows a fraction slightly bigger than $1/2$. Which one is it? Name a fraction that this strip might be representing.

Name two fractions that might be represented by strip C.

***Students could make a "sliding" ruler from construction paper, on which they could create various representations for fractions and pose questions to the class.***

***See Activity 17, which suggests another model to evaluate students' understanding of fractions.***

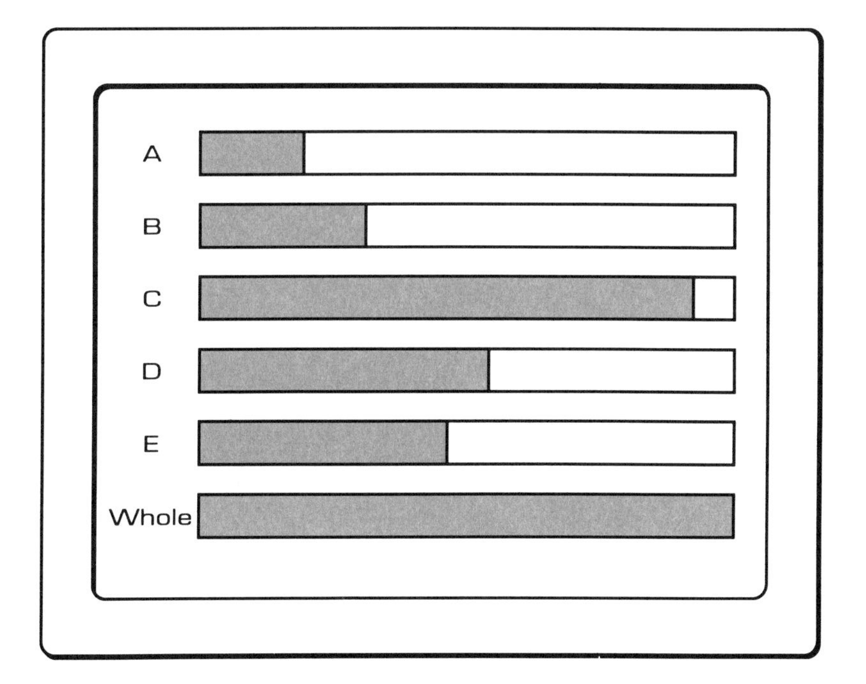

## ACTIVITY 20:
## FRACTION HUNT

**To the Teacher:** Generate a list of fractions such as those in the "Number Suspects" list. Give students clues until they guess the secret fraction. You may wish to have students generate their own list of clues for a secret number.

Use the clues to decide which number from the list of fractions is the secret number.

NUMBER SUSPECTS

$\frac{1}{2}$ $\frac{3}{4}$ $\frac{2}{5}$ $\frac{12}{11}$

$\frac{3}{5}$ $\frac{5}{8}$ $\frac{7}{10}$

Clue 1: I am greater than 0.5.

Clue 2: I am not equal to 0.75.

Clue 3: If you multiply me by 2, you get a number less than 2.

Clue 4: My denominator is a prime number.

Source: Woodward, Ernest. *The Math Detective.* Portland, Maine: J. Weston Walch, 1983. 

---

## ACTIVITY 21:
## SORTING FRACTIONS

**To the Teacher:** This activity can be used to analyze students' understanding of fractions. You may want to vary the activity by naming a fraction and having a student locate the appropriate region. Extend the activity by asking students to name fractions outside the three circles.

***As a follow-up question, you might ask students to tell which regions include an infinite number of fractions, which include a finite number of fractions, and which include an empty set. Encourage students to verify their thinking.***

Problem: Show the following diagram to students. Ask them to name a number that might be included in each region.

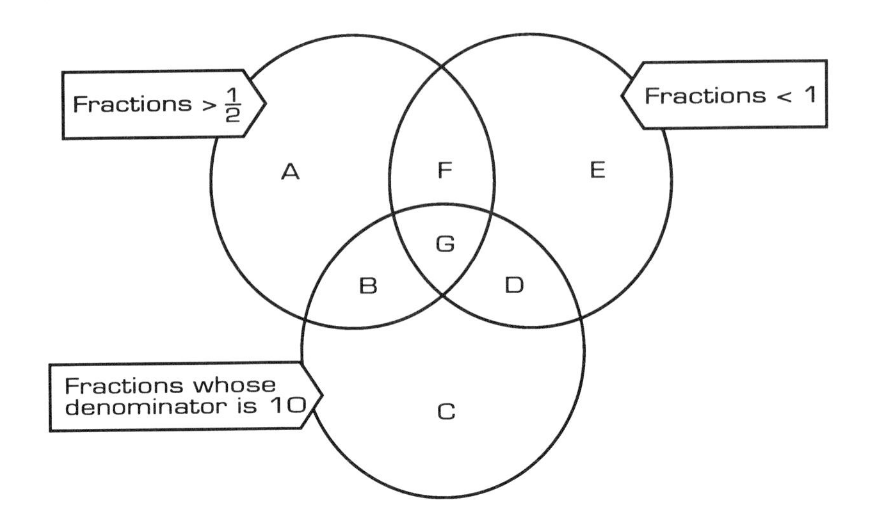

## ACTIVITY 22: EXPLORING NUMERATORS AND DENOMINATORS

**To the Teacher:** This activity focuses on students' understanding of fractions. Specifically, students are asked to make generalizations about fractions by comparing the numerator and the denominator.

*To get students started, have them consider specific values, for example, $a = 3$, $b = 2$, $c = 1$.*

*Use this activity to evaluate students' understanding by having them explain their thinking in written form.*

Suppose $a$, $b$, and $c$ represent whole numbers different from 0. Also, suppose that $a > b > c$. What can you say about each of these fractions?

$$\frac{a}{b} \qquad \frac{b}{a} \qquad \frac{b}{c}$$

$$\frac{c}{b} \qquad \frac{a}{c} \qquad \frac{c}{a}$$

If possible, tell which is larger. Justify your thinking.

$$\frac{a}{c} \text{ or } \frac{b}{c} \qquad \frac{a}{b} \text{ or } \frac{b}{b} \qquad \frac{a}{b} \text{ or } \frac{a}{c}$$

*Solutions: $a/b > 1$; $b/a < 1$; $b/c > 1$; $c/b < 1$; $a/c > 1$; $c/a < 1$; $a/c > b/c$; $a/b > b/b$; $a/b < a/c$.*

---

## ACTIVITY 23: ESTIMATING FRACTION SUMS

**To the Teacher:** This activity uses a number line to model answers for fraction addition problems. Encourage students to think about the size of each fraction relative to ½ in making their estimates.

Pose problems such as those listed below. Students are to estimate the answer to each by indicating the answer with an arrow on the number line as illustrated in the example.

*You might put the number line on the board and make "movable" arrows with tape on the back. Students would then tape their arrows on the number line as they describe their thinking.*

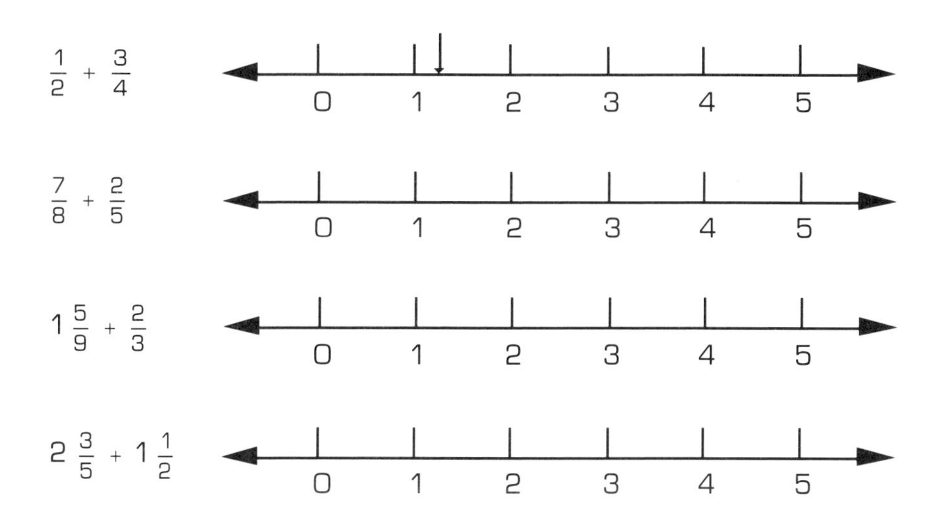


*One student might explain his thinking on the last problem in this way: "First, I'd find 2 3/5, then I'd move 1 whole unit right of it, then I'd move another half unit right, ending here."*

Source: Hoffer, Alan R., ed. *Mathematics Resource Project: Number Sense and Arithmetic Skills.* Palo Alto, Calif.: Creative Publications, 1978. Used with permission of the University of Oregon.

## ACTIVITY 24:
## OPERATIONS ON FRACTIONS

***See pages 11–13 for sample dialogue and a discussion of how to use this activity for evaluation purposes.***

**To the Teacher:** This activity is presented in the evaluation section of the *Curriculum and Evaluation Standards for School Mathematics* (NCTM 1989, p. 203). It is useful for clarifying students' understanding of the effects of certain operations on fractions.

Display the number line and ask students questions such as those indicated here and others that come to mind. Encourage students to justify their answers by explaining their reasoning.

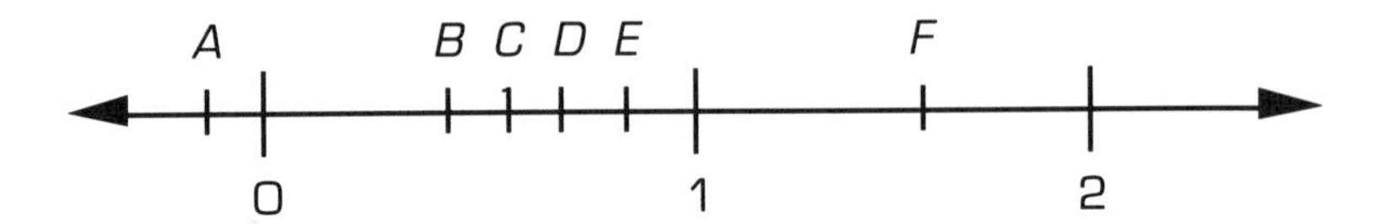


1. If the fractions represented by the points *D* and *E* are multiplied, what point on the number line best represents the product?
2. If the fractions represented by the points *C* and *D* are multiplied, what point on the number line best represents the product?
3. If the fractions represented by the points *B* and *F* are multiplied, what point on the number line best represents the product?
4. Suppose 20 is multiplied by the number represented by *E* on the number line. Estimate the product.
5. Suppose 20 is divided by the number represented by *E* on the number line. Estimate the quotient.

---

***You may want to provide calculators (four-function or fraction) so that students can quickly generate examples to verify their thinking.***

***Solutions: 1. <; 2. CT; 3. CT; 4. >; 5. CT; 6. <; 7. CT; 8. <. (Students will need to provide more than one example to justify their "can't tell" answers: one that includes a solution of > and one that requires the < symbol to complete the expression.)***

## ACTIVITY 25:
## MULTIPLYING AND DIVIDING FRACTIONS

**To the Teacher:** In this activity students are encouraged to make generalizations about operations involving fractions of various sizes.

Suppose that $a > 1$, $0 < b < 1$, and $0 < c < 2$. Ask students to complete each sentence by filling in the blank with <, =, >, or CT (for "can't tell").

1. $a \cdot b \; \square \; a$
2. $b \cdot c \; \square \; b$
3. $a \cdot b \cdot c \; \square \; b$
4. $a \div b \; \square \; a$
5. $a \div c \; \square \; a$
6. $b \div c \; \square \; b$
7. $c/c \; \square \; c$
8. $b^2 \; \square \; b$

♦ ♦ ♦ ♦ ♦ ♦ ♦ ♦

## ACTIVITY 26:
## EXPLORING TENTHS AND HUNDREDTHS

**To the Teacher:** Students can explore tenths and hundredths by using the constant addend feature of a calculator. By beginning at 0 and counting by tenths to 10, students gain an understanding of the relationship between decimals and whole numbers.

***Calculators vary slightly in the keystroking necessary to call on the constant addend feature. You may need to vary the suggested keystroking sequence (e.g., some calculators require that you enter the constant addend, then enter the + symbol twice).***

Have students key in 0 + 0.1 and then continue to press the equal (=) key, naming each number before it appears (e.g., "one-tenth, two-tenths, three-tenths," and so on). Ask them to notice what happens when they reach 0.9—what comes next? Many students will say "ten-tenths." Take this opportunity to highlight the different forms of ten-tenths (as a fraction and as its efficient decimal form of 1). Have students continue this counting aloud as they bridge several whole numbers. You may want to have someone record on the board the numbers named so that later, students can study the relative size of various decimals (e.g., Which is larger, 3.4 or 4.3?).

Next, ask students to count by hundredths (0.01) and to name each number displayed. Which is faster, counting by tenths to 10 or counting by hundredths to 1?

***Follow-up questions: How long will it take to count to 10 by tenths? To 100 by hundredths?***

---

## ACTIVITY 27:
## DECIMALS AND PLACE VALUE

**To the Teacher:** Generate problems similar to this one for further discussion. Students could also generate their own problems.

Problem: Some digits in this decimal number are hidden.

**____ 0. ____ 9**

Name the number if it is the largest number possible and has no two digits alike.

Name the number if it is the smallest number possible and has no two digits alike.

Name the number if it is between 40 and 50.

Name the number if it is about 40 ½.

♦ ♦ ♦ ♦ ♦ ♦ ♦ ♦

## ACTIVITY 28: ESTIMATING DECIMALS BETWEEN 0 AND 1

**To the Teacher:** This activity provides an opportunity for students to make estimates regarding decimals between 0 and 1.

Materials: Each pair of students needs a set of small markers (e.g., five to ten grains of rice), a number line, and an "accuracy ruler" as shown below.

Place students in pairs with the necessary material. In turn, each student asks his or her partner to locate a certain decimal; for example, "Show me 0.4" or "Show me 0.36." The partner estimates the location of the specified number along the number line and places a grain of rice as a marker. To check for accuracy, the other student uses the accuracy ruler. If the estimate is within one tenth of the actual location, the student making the estimate receives one point. The students now switch roles. The first person to score ten points is the winner.

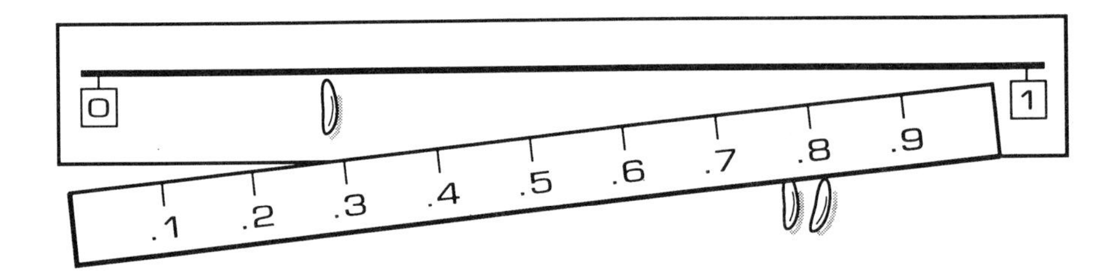


Source: Dougherty, Barbara J., and Joann E. Nelson. "Problem Solving, Estimation, and Number Sense with Rational Numbers." Paper presented at the annual meeting of the Greater San Diego Council of Teachers of Mathematics, San Diego, Calif., February 1989.

## ACTIVITY 29: ESTIMATING WITH DECIMALS

**To the Teacher:** Pose this type of question prior to any paper-and-pencil computation work. Encourage students to consider the bounds for each answer. For example, in the first problem, will the answer be more or less than 146? Will it be more or less than 73 (half of 146)?

Problem: Without calculating, tell all you know about the solution to these problems:

1. $146 \times 0.76$
2. $7.8 \times 0.98$
3. $45.1 \times 1.05$
4. $16 \div 0.5$
5. $39.5 \div 0.95$
6. $436.2 \div 0.63$
7. $82.5 \div 1.2$

***At first, do problems like these together in class. Later, let students write their answers in journal format so you can evaluate their individual understanding.***

---

## ACTIVITY 30: ESTIMATING DECIMAL PRODUCTS

**To the Teacher:** This activity is an extension of Activity 14, Estimating Whole Number Products. It focuses attention on the result of multiplying by "special" decimals (0.49, 0.9, 1.1, 1.8)—each being close to a common decimal or a whole number.

Four-Decimals-in-a-Row

Choose one member of the class to be the "calculator." Provide that student with a calculator. Divide the rest of the class into two teams. To begin, display the grid and the factor boards below. In turn, each team chooses two factors, one from the circular factor board and one from the square factor board. If the product of those numbers is displayed on the grid, the team captures that cell. The first team to capture four cells in a row (vertically, horizontally, or diagonally) is the winning team.

***Individual calculators are not needed for this activity. In fact, one overhead calculator is recommended for checking estimates.***

Grid

| | | | |
|---|---|---|---|
| 221.4 | 88.2 | 82.8 | 110.7 |
| 107.8 | 9 | 60.27 | 135.3 |
| 2.45 | 176.4 | 41.4 | 48.02 |
| 4.5 | 50.6 | 5.5 | 22.54 |

Factor boards

Square board: 46, 98, 123, 5

Circular board: 0.9, 1.8, 1.1, 0.49

## ACTIVITY 31:
## DECIMAL PLACEMENT

*Encourage students to estimate the value of the product by approximating each factor. For example, on question 2 a student might think, "It would be about half of 500, or 250. The decimal must go after the 1 to make the answer 291.357."*

*Note that all ending 0s have been eliminated from the decimal representation.*

*This activity could be used to evaluate students' estimation reasoning.*

**To the Teacher:** Students are asked to place the decimal point to indicate the correct solution. Encourage children to think holistically about each problem and to determine the placement of the decimal by what seems reasonable.

Each of the multiplication and division problems below has been carried out except for placing the decimal point. Place each decimal point using estimation.

1. 7.836 × 4.92 = 3 8 5 5 3 1 2
2. 534.6 × 0.545 = 2 9 1 3 5 7
3. 5.03 × 17.6 = 8 8 5 2 8
4. 49.05 × 6.044 = 2 9 6 4 5 8 2
5. 4.436 × 0.49 × 29.5 = 6 4 1 2 2 3 8
6. 68.64 ÷ 4.4 = 1 5 6
7. 400.14 ÷ 85.5 = 4 6 8
8. 0.735 ÷ 0.7 = 1 0 5
9. 51.1875 ÷ 1.05 = 4 8 7 5
10. 3.773 ÷ 0.98 = 3 8 5

---

## ACTIVITY 32:
## MULTIPLYING AND DIVIDING BY NUMBERS NEAR 1

**To the Teacher:** In this activity students are encouraged to fill in each blank by thinking about the effect of the indicated operation. All can be answered without calculation. Encourage students to verbalize their thinking for each problem. After completing all the problems, you might substitute ÷ for the × symbol in each problem. Ask students to reevaluate each sentence.

*To help students make and verbalize generalizations, ask questions such as the following: What happens when we multiply two numbers that are both greater than 1? What happens when we multiply a whole number by a fraction between 0 and 1? What happens when we multiply two numbers that are both between 0 and 1?*

Ask students to study each problem. Without performing any calculation they are to decide if $<$, $=$, or $>$ would complete each sentence. Ask them to justify their thinking.

1. 246 × 1.3 □ 246
2. 920 × 0.8 □ 920
3. 98 × 1.001 □ 98
4. $32 \times \frac{1}{2}$ □ 32
5. $\frac{1}{2} \times \frac{7}{8}$ □ $\frac{1}{2}$
6. $\frac{1}{2} \times \frac{7}{8}$ □ $\frac{7}{8}$

## ACTIVITY 33:
## EXPLORING FACTORS NEAR 1

**To the Teacher:** This game encourages students to think about the effect of multiplying when one factor is near 1.

Target Multiplication

To start this game, give students a target range and a starting value such as the following:

START: 36    TARGET RANGE: 2000–2100

One person in the class enters 36 into the calculator (an overhead calculator works very nicely) and presses ⊠. Another member of the class estimates a number that when multiplied by 36 will produce a product in the target range. For example, suppose a student estimates 50. The product (1800) misses the target range. Another student must now estimate a factor to multiply by the new number, 1800, to produce a product in the target range. Suppose a student estimates 1.2. The product of 1800 and 1.2 is 2160, which is just a bit too large.

Play continues in this manner until a student makes an estimate that produces a product in the target range. The "score" for the game is the number of guesses needed to hit the target range. The goal is to use as few guesses as possible.

***Create other start numbers and target ranges to form new games. Another way to extend the game is to use division as the operation rather than multiplication. For example, the target range might be 40–45 and the start value might be 2350.***

***This game first appeared in a book by George Immerzeel,*** **'77 Ideas for Using the Rockwell 18R in the Classroom,** ***published in 1976 by School Images, Inc. The book is no longer in print.***

## *ACTIVITY 34:*
## *MULTIPLICATION BY DECIMALS NEAR 0, ½, AND 1*

**To the Teacher:** This activity helps students explore the effect of multiplying by decimals near 0, ½, and 1. After students have generated and verified many of their own examples with a calculator, encourage them to make generalizations from the pattern that emerges.

What Happens When…?

What happens when you multiply by a number less than 1? Ask students to explore this question by completing the table.

| Pick a whole number | Multiply by 0.05 | Multiply by 0.48 | Multiply by 0.9 |
|---|---|---|---|
| | | | |
| | | | |
| | | | |
| | | | |

Questions to consider:

1. In general, what happens when you multiply a whole number by 0.05?
2. In general, what happens when you multiply a whole number by 0.48?
3. In general, what happens when you multiply a whole number by 0.9?

***Possible answers: 1. You get a result that is small compared to the other factor (or one-twentieth the size of the other factor).***
***2. You get a result that is a little less than half the size of the other factor.***
***3. You get a result that is a little less than the other factor (one-tenth less).***

## ACTIVITY 35:
## EXPLORING THE EFFECT OF OPERATIONS ON DECIMALS

**To the Teacher:** This activity provides an opportunity for students to explore the effect of addition, subtraction, multiplication, and division on decimal numbers. Encourage students to trace several paths through the maze while always looking for the path that will yield the greatest increase in the calculator's display. Note: Students often shy away from dividing by decimals less than 1, so you may want to discuss the general effect of dividing by a number less than 1 or multiplying by a number greater than 1.

Give each student a calculator and a copy of the MAZE. Students are to choose a path through the MAZE. To begin, have the students key in 100 on the calculator. For each segment chosen on the maze, the students should key in the assigned operation and number. The goal is to choose a path that results in the largest value at the finish of the maze. Students may not retrace a path or move upward in the maze.

***A template of the MAZE board can be found in Appendix 2.***

***Possible follow-up activities include finding the path that leads to the smallest finish number or finding a path that leads to a finish number as near the start number (100) as possible.***

***The largest possible finish number is about 6332.***

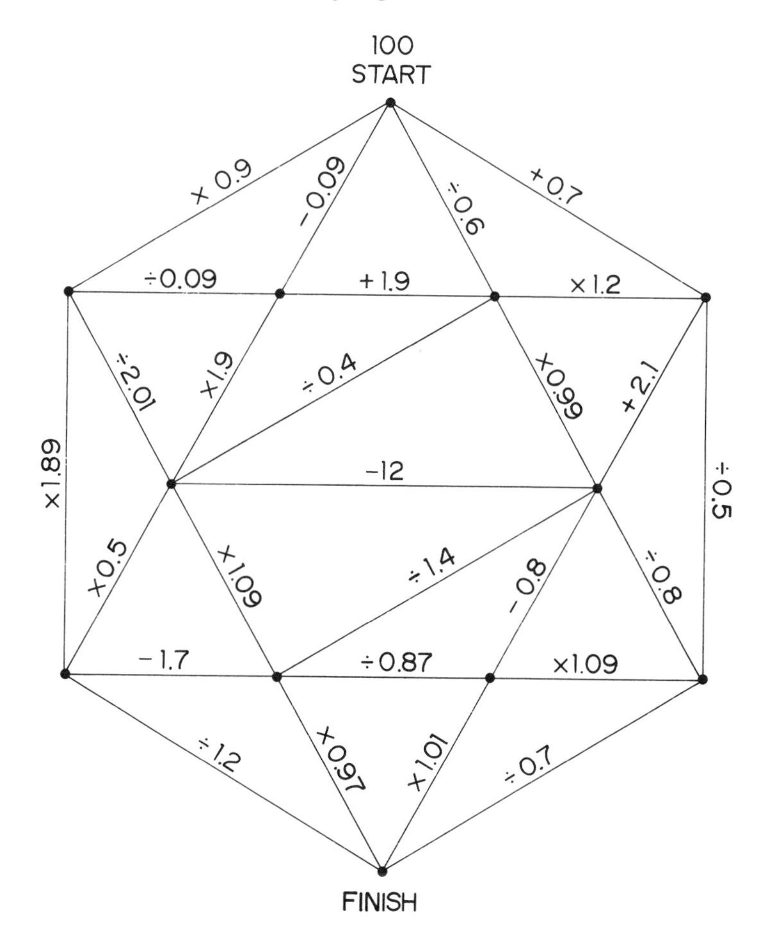


Source: Morris, Janet. *How to Develop Problem Solving Using a Calculator.* Reston, Va.: National Council of Teachers of Mathematics, 1981.

### ACTIVITY 36: EXPLORING EQUIVALENT EXPRESSIONS WITH DECIMALS

**To the Teacher:** In this activity students are asked to compare various expressions to the expression 12.8 × 48 and to choose those that are equivalent. As the students study each expression, encourage them to verbalize their thinking. As an extension, have students develop additional equivalent expressions.

***Students will likely think about each expression in different ways. For example, here is how three students verified that 64 × 9.6 is equivalent to 12.8 × 48:***

***"Multiply 48 by 4/3 to get 64, so you multiply 3/4 by 12.8 (9.6) to get an equivalent expression."***

***"Factor to get 6.4 × 2 × 6 × 8 or 6.4 × 12 × 8 or 6.4 × 96, which is the same as 64 × 9.6."***

***"Multiply one factor by 5 and divide the other by 5—64 ÷ 5 is 12.8 and 9.6 × 5 is 48. So they are equivalent."***

***As students share a variety of approaches, they gain a better appreciation that mathematics is a flexible tool for exploration rather than a fixed set of rules.***

***Answers: Expressions 1, 4, 5, 6, 7, and 8 are equivalent to 12.8 × 48.***

Which of these expressions have the same product as 12.8 × 48? Explain your thinking.

1. 128 × 4.8
2. 10.8 × 50
3. 128 × 4 × 0.8
4. 64 × 9.6
5. 2 × 32 × 9.6
6. 6.4 × 96
7. 25.6 × 24
8. 2 × 6.4 × 4 × 12

## ACTIVITY 37:
## PERCENT BENCHMARKS

**To the Teacher:** Students will likely benefit from discussion with fellow classmates as they complete this activity. You may want to organize them into small groups to encourage this discussion.

Ask students to read and reflect on each statement below and then to choose an answer from the answer list. The answers will vary; you may want students to continue this activity by researching each answer using appropriate sources of information or by conducting a survey.

Complete each statement below using one of the following choices:

0 percent
Fewer than 10 percent
About 25 percent
Fewer than 50 percent
About 50 percent
More than 50 percent
About 75 percent
At least 90 percent
100 percent

***If 0 percent is not used, ask students to give an example of a statement that would be answered with 0 percent. If 100 percent is not used, ask students to give an example of a statement that would be answered with 100 percent. Continue with other unused percentages.***

Statements:

1. ________________ of the students in my classroom are left-handed.
2. ________________ of the students in my classroom have red hair.
3. ________________ of the students in my school like hamburgers.
4. ________________ of the students in my school like baseball.
5. ________________ of the students in my school are wearing tennis shoes today.
6. ________________ of the people in my town are over ninety years old.
7. ________________ of the people in my town own a car.
8. ________________ of the people in my state are female.

## ACTIVITY 38: PERCENT OF A WHOLE

***Encourage students to trace over sections of the graph and to lay their tracings over other sections to estimate the size of each. Remind them that their total estimate for A, B, C, D, and E should be about 100 percent.***

**To the Teacher:** Give each student a copy of the graph illustrated below. You also might want to prepare a transparency of this diagram. Discuss with students which sections are easiest to estimate and which ones are more difficult.

Problem: Estimate the size (as a percent) of each section of the circle.

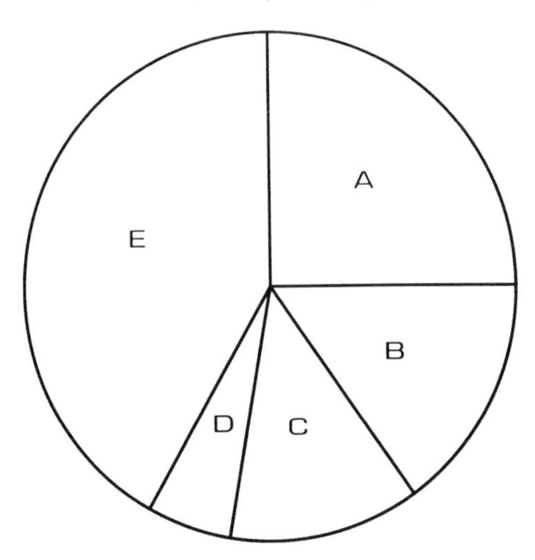


---

## ACTIVITY 39: EXPLORING EQUIVALENT EXPRESSIONS WITH PERCENTS

**To the Teacher:** In this activity students are asked to use their knowledge of percents to determine equivalent expressions and then to explain their choices. You might also ask students to generate other equivalent expressions.

In each row, choose the item that does not belong and explain your thinking.

| | | | | |
|---|---|---|---|---|
| 1. | $\frac{3}{4}$ | 0.75 | 0.34 | 75% |
| 2. | 20% of 40 | 40% of 20 | 10% of 80 | 5% of 20 |
| 3. | $3\frac{1}{4}$ | 0.325 | 3.25 | 325% |
| 4. | 50% of 18 | 18(0.5) | $\frac{1}{2}$% of 18 | 18% of 50 |
| 5. | $100 at 6% for 2 years | $100 at 8% for 1 year | $100 at 2% for 4 years | $100 at 4% for 2 years |
| 6. | percent change from 80 to 120 | percent change from 66 to 99 | percent change from 40 to 80 | percent change from 48 to 72 |

Source: Glatzer, David J., and Joyce Glatzer. *Math Connections.* Palo Alto, Calif.: Dale Seymour Publications, 1989.

## ACTIVITY 40: ESTIMATING PERCENTS

**To the Teacher:** In this activity students are asked to estimate the percentage of numbers between 1 and 100 that contain certain characteristics. You may want to display a number chart for verifying the estimate.

| 0 | 1 | 2 | 3 | 4 | 5 | 6 | 7 | 8 | 9 |
|---|---|---|---|---|---|---|---|---|---|
| 10 | 11 | 12 | 13 | 14 | 15 | 16 | 17 | 18 | 19 |
| 20 | 21 | 22 | 23 | 24 | 25 | 26 | 27 | 28 | 29 |
| 30 | 31 | 32 | 33 | 34 | 35 | 36 | 37 | 38 | 39 |
| 40 | 41 | 42 | 43 | 44 | 45 | 46 | 47 | 48 | 49 |
| 50 | 51 | 52 | 53 | 54 | 55 | 56 | 57 | 58 | 59 |
| 60 | 61 | 62 | 63 | 64 | 65 | 66 | 67 | 68 | 69 |
| 70 | 71 | 72 | 73 | 74 | 75 | 76 | 77 | 78 | 79 |
| 80 | 81 | 82 | 83 | 84 | 85 | 86 | 87 | 88 | 89 |
| 90 | 91 | 92 | 93 | 94 | 95 | 96 | 97 | 98 | 99 |

Ask students to estimate the percentage of numbers between 1 and 100 that—

1. contain only even digits;
2. contain only odd digits;
3. contain both even and odd digits;
4. contain the digit 2;
5. contain digits that add to 10;
6. are even;
7. are multiples of 3;
8. are composite.

Source: Cook, Marcy. "Ideas." *Arithmetic Teacher* 36 (November 1988): 31–36.

## *ACTIVITY 41: MEASURING ANGLES*

**To the Teacher:** When learning to use a protractor, students often become confused about which scale of the protractor to use when reading the angle measure. This activity offers an opportunity to remind students that angle measures, like answers to computation work, should make sense. In this context, benchmarks of 0-degree, 90-degree, and 180-degree angles help us determine the reasonableness of the measure.

90°

180° 0°

Before introducing the standard protractor, show students how to make their own crude protractors. Begin by cutting a half circle out of paper. Mark the center of the straight edge. Fold and crease the half circle to form a quarter circle. Open the fold and mark the 0-degree, 90-degree, and 180-degree points.

Indicate to students that they now have crude protractors, which allows them to estimate measures of angles. Provide a variety of angles for the students to "measure," including acute, right, and obtuse angles. Establish the need for other markings on the protractor to get a closer approximation of each angle. Then introduce and distribute standard protractors. Draw students' attention to the two scales, one extending from 0 to 180 degrees beginning at the right edge of the protractor and the other extending from 0 to 180 degrees beginning at the left edge of the protractor.

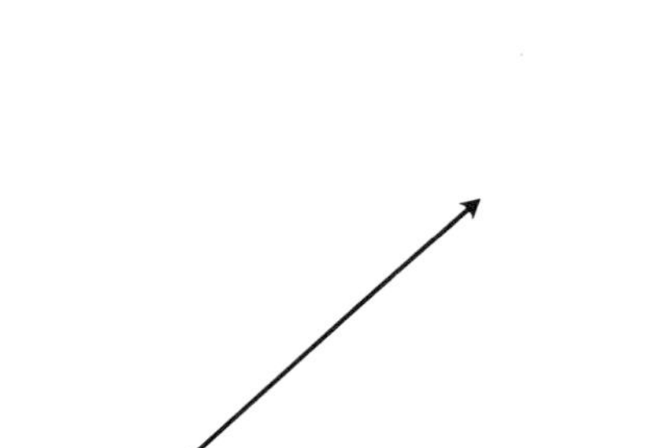

Ask students to look back at the angles whose measures they estimated earlier and to measure each using the standard protractor. Encourage students to decide which scale on the protractor to use by thinking about their earlier estimates and about what makes the most sense. For example, the angle shown here is acute and is therefore a 40-degree angle, not a 140-degree angle.

---

## *ACTIVITY 42: MEASUREMENT ESTIMATION*

***Encourage students to make estimates for each question before they launch the investigation.***

***Other topics to explore as extensions to this activity include water conservation and solid waste disposal. See* Mathematics and Global Survival *(Schwartz 1989) for ideas and data.***

**To the Teacher:** The questions included in this problem are open-ended and will likely generate exploration, research, and group discussion. You may want to investigate the first question as a whole-class activity, and then let students work in small groups to tackle some of the related questions.

Pose the following problem and ask groups of students to explore ways to answer each question.

Problem: How much water do you drink in one year? Enough to fill a bathtub? Enough to fill a swimming pool? How much water does your family use in one year (for drinking, washing, watering, and so on)? Enough to fill a swimming pool? Enough to fill a lake?

♦ ♦ ♦ ♦ ♦ ♦ ♦ ♦

## ACTIVITY 43: INTERPRETING GRAPHS

**To the Teacher:** This activity is designed to encourage students to make sense of line graphs as they evaluate information presented within the graph.

The graph presented below conveys the projected earnings of Tim and Rustin's hot dog stand. Their overhead costs are $35 plus 25 cents for each hot dog they sell to customers. The charge for a hot dog is 75 cents. Use the graph to answer each question. Justify your thinking.

1. Why are negative values displayed on the graph?

2. About how many hot dogs will the boys need to sell before they turn a profit?

3. "I'd like to make $100 profit on this deal," said Tim. How many hot dogs will the boys need to sell to reach the goal?

4. Describe the relationship between hot dogs sold and profit.

5. Describe how the graph would change if the boys charged more for each hot dog. How would it change if they charged less for each hot dog?

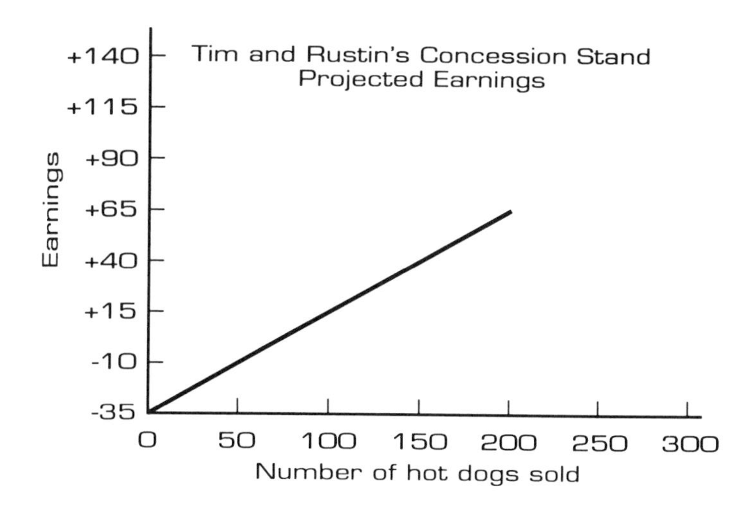

♦ ♦ ♦ ♦ ♦ ♦ ♦ ♦

## REFERENCES


Artzt, Alice F., and Claire M. Newman. *How to Use Cooperative Learning in the Mathematics Class.* Reston, Va.: National Council of Teachers of Mathematics, 1990.

Baroody, Arthur J. "One Point of View: Manipulatives Don't Come with Guarantees." *Arithmetic Teacher* 37 (October 1989): 4–5.

Bezuk, Nadine. *Understanding Rational Numbers and Proportion.* Reston, Va.: National Council of Teachers of Mathematics, forthcoming.

Brownell, William. "Psychological Considerations in the Learning and the Teaching of Arithmetic." In *The Place of Meaning in Mathematics Instruction: Selected Theoretical Papers of William A. Brownell.* Vol. 21. Edited by James F. Weaver and Jeremy Kilpatrick, pp. 19–51. Palo Alto, Calif.: Stanford University, School Mathematics Study Group, 1972.

Burns, Marilyn. "Beyond the Right Answer...Helping Your Students Make Sense out of Math." *Learning88* 16 (January 1988): 31–36.

Charles, Randall, Frank Lester, and Phares O'Daffer. *How to Evaluate Progress in Problem Solving.* Reston, Va.: National Council of Teachers of Mathematics, 1987.

Davidson, Neil, ed. *Cooperative Learning in Mathematics.* Menlo Park, Calif.: Addison-Wesley Publishing Co., 1989.

Geddes, Dorothy. *Geometry in the Middle Grades.* Reston, Va.: National Council of Teachers of Mathematics, 1992.

Glatzer, David J., and Joyce Glatzer. "No Answer Please." *Arithmetic Teacher* 36 (February 1989): 38–39.

Hirsch, Christian R., and Glenda Lappan. "Transition to High School Mathematics." *Mathematics Teacher* 82 (November 1989): 614–18.

Howden, Hilde. "Teaching Number Sense." *Arithmetic Teacher* 36 (February 1989): 6–11.

Immerzeel, George. *'77 Ideas for Using the Rockwell 18R in the Classroom.* Foxboro, Mass.: School Images, Inc., 1976.

National Council of Teachers of Mathematics. *An Agenda for Action.* Reston, Va.: The Council, 1980.

______. *Curriculum and Evaluation Standards for School Mathematics.* Reston, Va.: The Council, 1989.

______. Curriculum and Evaluation Standards for School Mathematics Addenda Series, Grades K–6, edited by Miriam A. Leiva. Reston, Va.: The Council, 1991–92.

______. Curriculum and Evaluation Standards for School Mathematics Addenda Series, Grades 5–8, edited by Frances R. Curcio. Reston, Va.: The Council, 1991–92.

______. Curriculum and Evaluation Standards for School Mathematics Addenda Series, Grades 9–12, edited by Christian R. Hirsch. Reston, Va.: The Council, 1991–92.

______. *Professional Standards for Teaching Mathematics.* Reston, Va.: The Council, 1991.

National Research Council. *Everybody Counts: A Report to the Nation on the Future of Mathematics Education.* Washington, D.C.: National Academy Press, 1989.

Parker, Tom. *In One Day.* Boston: Houghton Mifflin, 1984.

Phillips, Elizabeth. *Patterns and Functions.* Reston, Va: National Council of Teachers of Mathematics, 1991.

Rowan, Thomas. Report of the Task Force on Addenda to the NCTM K–12 *Curriculum and Evaluation Standards for School Mathematics.* Unpublished report, November 1988.

Schwartz, Richard H. *Mathematics and Global Survival.* Needham Heights, Mass.: Ginn Press, 1989.

Sowder, Judith T., and Bonnie P. Schappelle, eds. *Establishing Foundations for Research on Number Sense and Related Topics: Report of a Conference.* San Diego: San Diego State University, Center for Research in Mathematics and Science Education, 1989.

Steen, Lynn Arthur. "Teaching Mathematics for Tomorrow's World." *Educational Leadership* 47 (September 1989): 18–22.

Thornton, Carol A., and Sally C. Tucker. "Lesson Planning: The Key to Developing Number Sense." *Arithmetic Teacher* 36 (February 1989): 18–21.

Vygotsky, Lev. *Thought and Language.* Cambridge, Mass.: MIT Press, 1934/1986.

Whitin, David J. "Number Sense and the Importance of Asking 'Why?'" *Arithmetic Teacher* 36 (February 1989): 26–29.

*The World Almanac and Book of Facts, 1990.* New York: World Almanac, 1990.

Zawojewski, Judith. *Dealing with Data and Chance.* Reston, Va.: National Council of Teachers of Mathematics, 1991.

♦ ♦ ♦ ♦ ♦ ♦ ♦ ♦

## *RECOMMENDED READING*

### *Books*


Brownell, William. "When Is Arithmetic Meaningful?" In *The Place of Meaning in Mathematics Instruction: Selected Theoretical Papers of William A. Brownell.* Vol. 21. Edited by James F. Weaver and Jeremy Kilpatrick, pp. 73–90. Palo Alto, Calif.: Stanford University, School Mathematics Study Group, 1972.

Hope, Jack A. *Numeracy: A Common Essential Learning for Saskatchewan Students.* Monograph, Department of Curriculum Studies, University of Saskatchewan, Saskatoon, Saskatchewan, 1987.

Paulos, John Allen. *Innumeracy: Mathematical Illiteracy and its Consequences.* New York: Hill and Wang, 1988.


### *Articles*


Burnett, Peg H. "A Million! How Much Is That?" *Arithmetic Teacher* 29 (September 1987): 49–50.

Kennedy, Bill. "Writing Letters to Learn Math." *Learning* 13 (February 1985): 59–61.

Leutzinger, Larry P., and Myrna Bertheau. "Making Sense of Numbers." In *New Directions for Elementary School Mathematics*, pp. 111–122. Edited by Paul R. Trafton and Albert P. Shulte. Reston, Va.: National Council of Teachers of Mathematics, 1989.

National Council of Teachers of Mathematics. *Arithmetic Teacher* 36 (February 1989). Focus issue: Number Sense.

Philipp, Randolph A. "Piano Tuners and Problem Solving." *Mathematics Teacher* 82 (April 1989): 248–49.

Reys, Robert E., and Barbara J. Reys, eds. "Estimation and Mental Computation." *Arithmetic Teacher* 34 (September 1986–May 1987).

Turner, Rebecca. "Understanding Size." *Learning88* 16 (March 1988): 89.


### *Activity Books*


Burns, Marilyn. *A Collection of Math Lessons, Grades 3–6.* New Rochelle, N.Y.: Cuisenaire Company of America, Inc., 1987.

———. *The Good Time Math Event Book.* Palo Alto, Calif.: Creative Publications, 1977.

———. *The I Hate Mathematics! Book.* Boston: Little, Brown & Co., 1975.

———. *Math for Smarty Pants.* Boston: Little, Brown & Co., 1982.

Hope, Jack, Barbara Reys, and Robert Reys. *Mental Math in the Middle Grades.* Palo Alto, Calif.: Dale Seymour Publications, 1987.

Landwehr, James M., and Ann Watkins. *Exploring Data.* Palo Alto, Calif.: Dale Seymour Publications, 1986.

Newman, Claire M., Thomas E. Obremski, and Richard L. Scheaffer. *Exploring Probability.* Palo Alto, Calif.: Dale Seymour Publications, 1987.

Reys, Robert, Paul Trafton, Barbara Reys, and Judy Zawojewski. *Computational Estimation, Grades 6, 7, 8.* Palo Alto, Calif.: Dale Seymour Publications, 1987.

Reys, Robert, and Barbara Reys. *GUESS: A Guide to Using Estimation Skills and Strategies.* Palo Alto, Calif.: Dale Seymour Publications, 1983.

Seymour, Dale. *Developing Skills in Estimation, Books A and B.* Palo Alto, Calif.: Dale Seymour Publications, 1980.

Stenmark, Jean Kerr, Virginia Thompson, and Ruth Cossey. *Family Math.* Berkeley, Calif.: University of California, Lawrence Hall of Science, 1986.


### *Resource Books*


Anno, Mitsumasa. *Anno's Mysterious Multiplying Jar.* New York: Philomel, 1983.

Education Development Center. *Bicycles.* Pleasantville, N.Y.: Sunburst Communications, 1985.

———. *Ice Cream.* Pleasantville, N.Y.: Sunburst Communications, 1985.

Froman, Robert. *Less Than Nothing Is Really Something.* New York: Thomas Y. Crowell, 1973.

Parker, Tom. *In One Day.* Boston: Houghton Mifflin Co., 1985.

Schwartz, David M. *How Much Is a Million?* New York: Lothrop, Lee & Shepard Books, 1985.

———. *If You Made a Million.* New York: Lothrop, Lee & Shepard Books, 1989.

Appendix 1

## *EVALUATION*

Diana Lambdin Kroll and Frank K. Lester, Jr.

Teaching according to the vision of the *Curriculum and Evaluation Standards for School Mathematics* involves numerous changes in content and instruction, but it may also require a change in how you evaluate students' work. Evaluation includes much more than marking right and wrong answers. A number of alternative evaluation techniques are discussed in the following pages. But first, we reflect on the goals of evaluation.

### *WHY DO WE EVALUATE?*

In the past, evaluation was often thought of as synonymous with testing (and results of evaluation were frequently used primarily for grading); the *Curriculum and Evaluation Standards* recommends a broader conception. There are at least four reasons for collecting evaluation information:

- To make decisions about the content and methods of mathematics instruction
- To make decisions about classroom climate
- To help in communicating what is important
- To assign grades

#### *Making Decisions about Mathematics Instruction*

One of the most important reasons for gathering evaluation data is to make instructional decisions about the content of your mathematics program and the teaching methods you use. Data from classroom observations, from students' writing about mathematics, and from analyses of their written work can be used to diagnose students' strengths and weaknesses. Clearly, the precision with which you can diagnose strengths and weaknesses is highly influenced by the goal of the evaluation and the type of evaluation technique you use. The more types of evaluation data you can collect on your students, the better able you'll be to modify your instruction to meet their needs.

#### *Making Decisions about Classroom Climate*

A classroom climate conducive to having students actively involved in doing mathematics is essential to meeting the goals outlined in the *Curriculum and Evaluation Standards.* Conditions that contribute to classroom climate include the following:

- The commitment and enthusiasm exhibited by the teacher

  *Does the teacher convey to students that she enjoys teaching mathematics?*

- The frequency of lessons focused on doing mathematics as opposed to practicing skills

  *Are problem solving and discovery integral parts of the class?*

- The evaluation practice used

  *Are students given opportunities to explore and experiment without being graded?*

  *Is more than the answer evaluated?*

Perhaps the most important indicators of an appropriate classroom climate are students' attitudes and beliefs. Assessment techniques such as student self-inventories, student self-reports, and interviews or observations of students can provide the data you need to make judgments concerning the attitudes and beliefs about mathematics that are being fostered in your classroom.

#### *Communicating What Is Important*

As any teacher knows, students are experts at detecting what teachers consider important and unimportant. If you assign homework, be sure to hold students responsible for it. If you expect students to show all their work, be sure to look at more than just final answers. In general, students internalize as important those aspects of instruction that their teacher emphasizes and assesses regularly. The *Curriculum and Evaluation Standards* emphasizes problem solving, reasoning, communication, and making mathematical connections as important goals for any mathematics class. Your students will value progress in these areas only if your evaluation techniques indicate that you do.

#### *Assigning Grades*

A final reason for collecting evaluation data is to assign grades. It is important to understand that evaluation is not synonymous with grading. But, when you do decide to assign grades, the following guidelines may be useful:

- Advise students in advance when their mathematics work will be graded.
- Use a grading system that considers the process used to solve problems, not just the answer.
- Be aware that pupils may not perform as well when they are graded.
- Use as much evaluation data and as many different techniques as possible to help you assign grades.
- Consider using a testing format that matches your instructional format. (For example, if your students usually work in cooperative groups, consider testing performance in cooperative groups.)

### *AN OVERVIEW OF FOUR METHODS OF EVALUATION*

In the next few pages, we discuss a variety of evaluation techniques, and we comment on their usefulness in accomplishing the goals we've just discussed. The evaluation techniques are organized into four major categories: (1) *observing and questioning students*, (2) *using assessment data reported by students*, (3) *assessing students' written mathematics work*, and (4) *using multiple-choice or short-answer tests*. (Note: Each category of techniques is discussed in considerably more detail in Charles, Lester, and O'Daffer [1987].)

#### *Observing and Questioning Students*

Observing and questioning students while they are engaged in mathematics activities can provide invaluable information not only about their skills, but also about their thinking processes, their attitudes, and their beliefs. Such observation can be done either in a rather informal way, as you circulate around the room, or in a more structured and formal interview.

You can learn a lot about students by circulating unobtrusively as they work in small groups solving problems, and by interjecting questions to clarify your observations. But gathering evaluation data using classroom observation can be difficult to do because there is so much going on when students are working on problems. It is important to think ahead about what you want to evaluate, so that you can focus attention on the relevant aspects of students' work.

Similarly, it is important to think carefully about the types of questions you ask when you are observing. In everyday classroom situations, teachers ask students questions for a variety of reasons: to stimulate thinking, to provide hints, to test, or to demonstrate to other students what their peers know. When you are "wearing your evaluator's cap," you may need to suspend temporarily your natural tendency to teach. Instead, you will try to phrase your questions so that students' answers will give you information about understandings or behaviors you are trying to evaluate.

Record your observation findings briefly, objectively, and promptly. Depending on the purpose of the observation, you may choose to use a comment card, a checklist, a rating scale, or a journal for recording observations. Observation is probably the best way to get a firsthand look at how your students think about and do mathematics.

***Using Assessment Data Reported by Students***

Three techniques for collecting written data directly from students are self-reports, student self-inventories, and individual mathematics portfolios. The former two are especially appropriate for assessing students' feelings and beliefs about mathematics and for obtaining information about how they organize and monitor their work. The usefulness of such data depends, of course, on how accurately, completely, and candidly students report their actions, feelings, beliefs, and intentions. Individual portfolios, in contrast, provide a broader picture of a student's progress in mathematics over a grading period or a year.

*Student reports.* In student self-reports, ask your students to write or to dictate into a tape recorder a retrospective account of a just-completed problem-solving experience. Provide some structure for the report by asking students to respond to questions that focus on selected aspects of the problem and experience. In general, student reports are more frequently used to assess attitudes or beliefs than performance. And they certainly are not appropriate for grading purposes, since such use might affect the candidness of the reports.

*Student self-inventories.* Another, more structured type of student self-report is the student self-inventory. An inventory is a list of items, furnished by the teacher, to which the student responds selectively (either by simply checking those items that apply, or by marking, from a range of options, the degree to which the item applies). The most familiar type of inventory is an attitude or belief survey, but inventories can also be designed to gather information about students' evaluation of their own mathematical thinking and expertise; about students' preferences for, or familiarity with, particular types of problems; or about students' monitoring behaviors during problem solving. Because the information gathered from self-inventories is subjective and can be incomplete, you should probably use self-inventories only in combination with other evaluation techniques.

*Individual mathematics portfolios.* Like a professional artist's portfolio, an individual mathematics portfolio is more than just a folder of work. It is a collection of documents carefully selected by the student to provide a broad view of his or her range of interests and abilities in mathematics. Items in a portfolio will range from test papers to written self-reflections, from samples of homework to original student-written problems. It is important to limit the number of items that may be included in a portfolio so that students are forced to reflect on which samples of their work to include.

#### Assessing Students' Mathematics Work

When teachers grade students' written work, they usually first look at answers and then probe deeper for sources of errors if the answers are incorrect. Evaluating written work according to the spirit of the *Curriculum and Evaluation Standards* is similar, although more emphasis is generally given to process (how the student approaches the problem) and less to product (what answer is obtained). Clearly, the most difficult part of evaluating students' written work is deciding how to assess work that shows various types of errors. In this section we describe two specific methods for evaluating students' written work on problems: *analytic scoring* and *holistic scoring.*

*Analytic scoring.* Analytic scoring is particularly useful in assessing students' problem-solving efforts. Analytic scoring involves using a scale to assign points to certain phases of the mathematical problem-solving process. First identify the problem-solving phases that you want to evaluate, and then specify a range of scores to be awarded for various levels of performance in each phase. For example, in evaluating students' written work on word problems, your analytic scoring scale might assign four points to understanding the problem, four points to planning a solution, and two points to getting a correct answer. An analytic scoring scheme generally also includes specific criteria for awarding partial credit for each phase.

Analytic scoring methods are most appropriate for giving students feedback about their performance in key categories associated with mathematical problem solving, for obtaining diagnostic information about students' specific strengths and weaknesses, or for identifying specific aspects of mathematics that may require additional instructional time.

*Holistic scoring.* Unlike analytic scoring, which produces several numeric scores for each problem (each associated with a different aspect of problem-solving work), holistic scoring produces a single number assigned according to specific criteria related to the thinking processes involved in solving the problem. Holistic scoring is *holistic* because it focuses on the total solution (neither on the answer alone, nor separately on various aspects of the solution). Holistic scoring is most appropriate when you need a relatively quick, yet consistent, evaluation technique. For classroom use, focused holistic scoring should be used in combination with other, more informative evaluation techniques.

#### Using Multiple-Choice and Short-Answer Tests

Multiple-choice and short-answer tests are expedient (easy to administer and to score). Yet, such tests are actually the least satisfactory means of evaluating students' progress in mathematics. Although the *Curriculum and Evaluation Standards* makes it clear that there is much more to learning mathematics than producing correct answers, items on such tests usually focus only on whether students are able to find correct answers. In the classroom—where other, preferable methods of evaluation are possible—teachers should not spend their time attempting to construct and use multiple-choice or completion tests for assessing understanding in mathematics.

### CHOOSING EVALUATION TECHNIQUES

Clearly, your choice of evaluation technique needs to be based on a multitude of factors, such as the type of mathematical skill you are assessing, the number of students you need to evaluate and the amount of time available, your experience in teaching and evaluating higher-order thinking, the reason for your evaluation, and the availability of evaluation materials. Using several techniques enables you to evaluate students much more comprehensively. You might collect a variety of documents in a folder for each student (e.g., notes from classroom observations, copies of classwork or homework, lists of test grades, samples of self-assessment responses, and written essays), or you might have students submit their own mathematics portfolios. Your choice of evaluation techniques will be quite personal. But in the end, you need to choose techniques that are feasible for use in your particular classroom situation and that provide information appropriate for the goals of your evaluation.

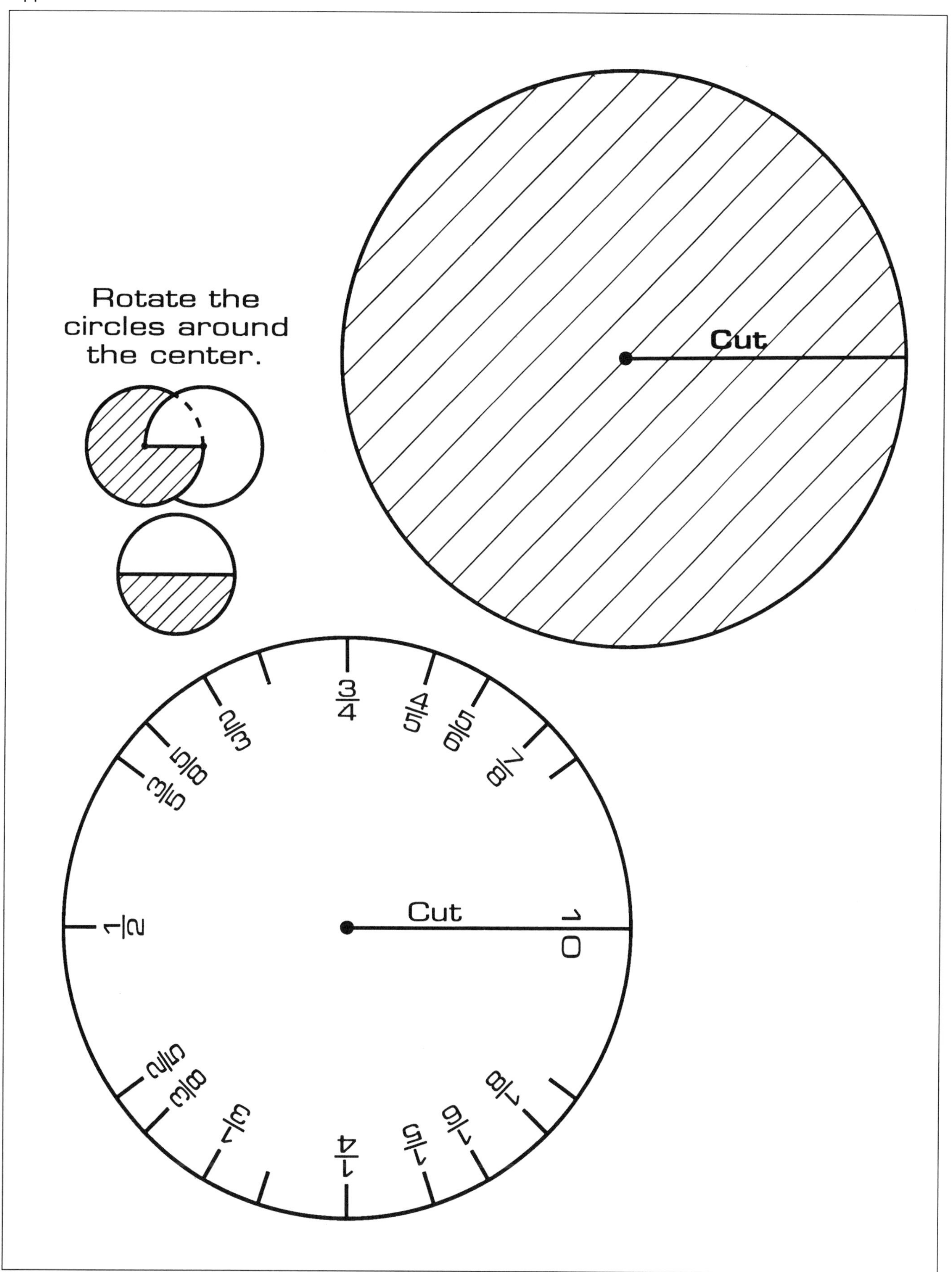
Rotate the circles around the center.
Cut
Cut
3/4
4/5
5/6
7/8
2/3
5/8
3/5
1/2
1/10
2/5
3/8
1/3
1/4
1/5
1/6
1/8

# Maze

## Playing Board

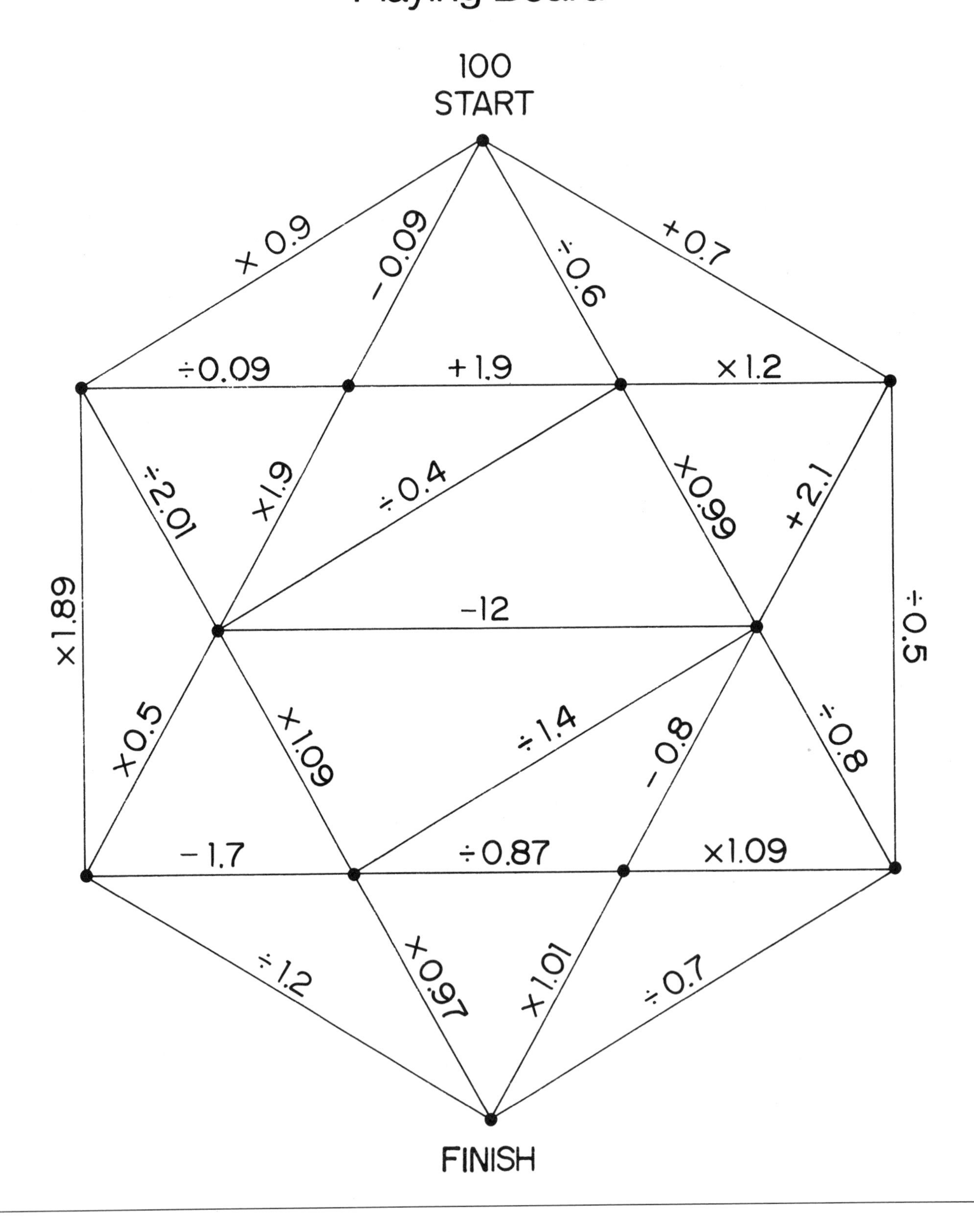